Ernst Schering Research Foundation Workshop 60
Stem Cells in Reproduction and in the Brain

Ernst Schering Research Foundation
Workshop 60

Stem Cells in Reproduction and in the Brain

J. Morser, S.-I. Nishikawa, H.R. Schöler
Editors

With 40 Figures

 Springer

Series Editors: G. Stock and M. Lessl

Library of Congress Control Number: 2006923356

ISSN 0947-6075
ISBN-10 3-540-31436-9 Springer Berlin Heidelberg New York
ISBN-13 978-3-540-31436-3 Springer Berlin Heidelberg New York

Springer is a part of Springer Science+Business Media
springer.com

Editor: Dr. Ute Heilmann, Heidelberg
Desk Editor: Wilma McHugh, Heidelberg
Production Editor: Monika Riepl, Leipzig
Cover design: design & production, Heidelberg
Typesetting and production: LE-TEX Jelonek, Schmidt & Vöckler GbR, Leipzig
21/3100/YL – 5 4 3 2 1 0   Printed on acid-free paper

Preface

Almost every day the media report on the latest advances and break-throughs in stem cell biology and regenerative medicine. These stories usually focus on either the short-term expectations for therapy or on the ethical issues posed by the new technologies. Within the scientific community, interest in stem cells has been increasing rapidly because the study of stem cells is generating new insights into the fundamental

processes of development and repair. In addition there is the possibility that use of stem cells could lead to significant improvements for patients with many chronic debilitating diseases.

The practical use of adult stem cells depends on the ability to mobilize them from their state of hibernation in the tissues and perhaps also to convert them from one lineage into another. In contrast, embryonic stem cells need to have their differentiation directed towards the desired cell type while ensuring that any potential for tumorigenesis is removed. In all cases, issues of safety have to be central to the development of therapeutics based on stem cells.

As part of Germany in Japan year, the Ernst Schering Research Foundation, the Max Planck Society and the Riken Center on Developmental Biology jointly organized a workshop on Stem Cells in Reproduction and in the Brain, which took place in Kobe, Japan on September 1st–3rd 2005. The purpose of the workshop was to discuss the present state of knowledge and future directions in this important field. Leading basic scientists and clinicians reviewed and discussed several timely topics within these main themes: (1) Stem cell properties (2) Transdifferentiation and fusion of stem cells (3) Identification of stem cells (4) Stem cells in reproduction and (5) Neural stem cells.

In both the reproductive system and in the brain, recent data from stem cell experiments has led to a reexamination of long held views. The brain contains neural stem cells with the ability to proliferate, migrate and differentiate throughout adult life. Germ cells were believed to be unique in developing separately from all other cells in the body. The possibility of using other cells to derive cells in lineages leading to sperm and oocytes has been shown from both embryonic stem cells and from bone marrow.

This book contains the proceedings of the workshop giving a summary of the basic science insights and the steps needed for converting those insights into practical therapies in the clinic. In addition to presenting the current thinking on the role of stem cells in reproduction and the brain, some of the basic unanswered questions and controversies are described.

John Morser
Shin-Ichi Nishikawa
Hans Schoeler

Contents

1   What Is a Stem Cell Niche?
     S.-I. Nishikawa, M. Osawa   1

2   New Strategy for Comprehensive Analysis
     of Gene Functions in Embryonic Stem cells
     J. Takeda, K. Horie, K. Yusa   15

3   Paternal Dual Barrier by Ifg2-H19 and Dlk1-Gtl2
     to Parthenogenesis in Mice
     T. Kono, M. Kawahara, Q. Wu, H. Hiura, Y. Obata   23

4   Cell-Cell Fusion as a Means to Establish Pluripotency
     J.T. Do, H.R. Schöler   35

5   Toward Reprogramming Cells to Pluripotency
     P. Collas, C.K. Taranger   47

6   Molecular Switches and Developmental Potential
     of Adult Stem Cells
     M. Zenke, T. Hieronymus   69

7   Adult Small Intestinal Stem Cells: Identification,
     Location, Characteristics, and Clinical Applications
     C.S. Potten, J.R. Ellis   81

8   Tracking Stem Cells In Vivo
     R. Yoneyama, E.R. Chemaly, R.J. Hajjar   99

9   Establishment of Nuclear Transfer
    Embryonic Stem Cell Lines from Adult Somatic Cells
    by Nuclear Transfer and Its Application
    T. Wakayama   111

10 Derivation of Germ Cells from Embryonic Stem Cells
    J. Kehler, K. Hübner, H.R. Schöler   125

11 Germline Recruitment in Mice:
    A Genetic Program for Epigenetic Reprogramming
    Y. Ohinata, Y. Seki, B. Payer, D. O'Carroll, M.A. Surani,
    M. Saitou   143

12 Transplantation of Germ Line Stem Cells for the Study
    and Manipulation of Spermatogenesis
    I. Dobrinski   175

13 Progenitor Cell-Based Myelination as a Model
    for Cell-Based Therapy of the Central Nervous System
    S.A. Goldman, J. Lang, N. Roy, S.J. Schanz, F.S. Sim, S. Wang,
    V. Washco, M.S. Windrem   195

14 Adult Neural Stem Cells
    and Central Nervous System Repair
    H. Okano   215

15 Stem Cell Therapy for Parkinson's Disease
    J. Takahashi   229

Previous Volumes Published in This Series   245

List of Editors and Contributors

Editors

Morser, J.
Nihon Schering K.K. Research Center,
BMA 3F, 1-5-5 Minatojima-minai-machi, Chuo-ku, Kobe 650-0047, Japan
(e-mail: john_morser@berlex.com)

Nishikawa, S.-I.
Center for Developmental Biology,
2-2-3 Minatojima minami-machi, Chuo-ku, Kobe 650-0047, Japan
(e-mail: nishikawa@cdb.riken.jp)

Schöler, H.R.
Max Planck Institute for Molecular Biomedicine,
Mendelstr. 7, 48149 Münster, Germany
(e-mail: schoeler@mpi-muenster.mpg.de)

Contributors

Chemaly, E.R.
Cardiovascular Research Center, Massachusetts General Hospital,
149 13th Street, Room 4215, Charlestown, MA 02129, USA

Collas, P.
Institute of Basic Medical Sciences, Department of Biochemistry,
University of Oslo,
P.O. Box, 1112 Blindern, 0317 Oslo, Norway
(e-mail: phillippe.collas@medisin.uio.no)

Do, J.T.
Max Planck Institute for Molecular Biomedicine,
Department of Cell and Developmental Biology,
Mendelstr. 7, 48149 Münster, Germany
(e-mail: jeong.tae.do@mpi-muenster.mpg.de)

Dobrinski, I.
Center for Animal Transgenesis and Germ Cell Research,
School of Veterinary Medicine, University of Pennsylvania,
Kennett Square, PA 19348, USA
(e-mail: dobrinski@vet.upenn.edu)

Ellis, J.R.
EpiStem Limited, Incubator Building,
Grafton Street, Manchester M13 9XX, UK
(e-mail: j.ellis@epistem.co.uk)

Goldman, S.
Division of Cell and Gene Therapy,
Departments of Neurology and Neurosurgery,
University of Rochester Medical Center,
601 Elmwood Avenue, Box 645, Rochester, NY 14642, USA
(e-mail: steven_goldman@urmc.rochester.edu)

Hajjar, R.
Cardiovascular Research Center, Massachusetts General Hospital,
149 13th Street, Room 4215, Charlestown, MA 02129, USA
(e-mail: rhajjar@partners.org)

Horie, K.
Department of Social and Environmental Medicine,
Graduate School of Medicine, Osaka University,
2-2 Yamadaoka, Suita, Osaka 565-0871, Japan

Hieronymus, T.
Institute for Biomedical Engineering,
Department of Cell Biology, Aachen University Medical School,
Pauwelsstrasse 20, 52074 Aachen, Germany

Hiura, H.
Department of BioScience, Tokyo University of Agriculture,
Setagaya-ku, Tokyo 156-8502, Japan

Hübner, K.
Max Planck Institute for Molecular Biomedicine,
Department of Cell and Developmental Biology,
Mendelstr. 7, 48149 Münster, Germany

Kawahara, M.
Department of BioScience, Tokyo University of Agriculture,
Setagaya-ku, Tokyo 156-8502, Japan

Kehler, J.
Germline Development Group, University of Pennsylvania,
School of Veterinary Medicine,
Center for Animal Transgenesis and Germ Cell Research,
382 West Street Road, Kennett Square, PA 19348, USA

Kono, T.
Department of BioScience, Tokyo, University of Agriculture,
Setagaya-ku, Tokyo 156-8502, Japan
(e-mail: tomohiro@nodai.ac.jp)

Lang, J.
Division of Cell and Gene Therapy,
Departments of Neurology and Neurosurgery,
University of Rochester Medical Center,
601 Elmwood Avenue, Box 645, Rochester, NY 14642, USA

Obata, Y.
Department of BioScience, Tokyo University of Agriculture,
Setagaya-ku, Tokyo 156–8502, Japan

O'Carrol, D.
Laboratory for Lymphocyte Signaling, the Rockefeller University,
1230 York Avenue, New York, NY 10021, USA

Ohinata, Y.
Riken Center for Developmental Biology,
2-2-3 Minatojima-minamachi, Chuo-ku, Kobe 650-0047 Japan

Okano, H.
Department of Physiology,
Keio University School of Medicine, Tokyo,
35 Shinanomachi, Shinjuku-ku, 160-8582, Japan
(e-mail: hidokano@sc.itc.keio.ac.jp)

Osawa, M.
Center for Developmental Biology,
2-2-3 Minatojima minami-machi, Chuo-ku, Kobe 650-0047, Japan
(e-mail: mosawa@cdb.riken.jp)

Payer, B.
Wellcome Trust/Cancer Research,
UK Gurdon Institute of Cancer and Developmental Biology,
University of Cambridge, Tennis Court Road, Cambridge, CB2 1QN, UK

Potten, C.S.
EpiStem Limited, Incubator Building,
Grafton Street, Manchester M13 9XX, UK
(e-mail: c.potten@epistem.co.uk)

Roy, N.S.
Weill Medical College of Cornell University,
425 East 61st Street, New York, NY 10021, USA

Saitou, M.
Riken Center for Developmental Biology,
2-2-3 Minimatojima-Minamimachi, Chuo-ku, Kobe 650-0047 Japan
(e-mail: saitou@cdb.riken.jp)

Schanz, J.
Division of Cell and Gene Therapy,
Departments of Neurology and Neurosurgery,
University of Rochester Medical Center,
601 Elmwood Avenue, Box 645, Rochester, NY 14642, USA

Seki, Y.
Riken Center for Developmental Biology,
2-2-3 Minatojima-minamachi, Chuo-ku, Kobe 650-0047 Japan

Sim, F.J.
Division of Cell and Gene Therapy,
Departments of Neurology and Neurosurgery,
University of Rochester Medical Center,
601 Elmwood Avenue, Box 645, Rochester, NY 14642, USA

Surani, M.A.
Wellcome Trust/Cancer Research,
UK Gurdon Institute of Cancer and Developmental Biology,
University of Cambridge, Tennis Court Road, Cambridge, CB2 1QN, UK

Takahashi, J.
Department of Neurosurgery, Kyoto University Graduate School of Medicine,
54 Kawahara-cho-Shogoin, Sakyo-ku, Kyoto 606-8507, Japan
(e-mail: jbtaka@kuhp.kyoto-u.ac.jp)

Takeda, J.
Department of Social and Environmental Medicine,
Graduate School of Medicine, Osaka University,
2-2 Yamadaoka, Suita, Osaka 565-0871, Japan
(e-mail: takeda@mr-envi-med.osaka-u.ac.jp)

Taranger, C.K.
Institute of Basic Medical Sciences, Department of Biochemistry,
University of Oslo, PO Box, 112 Blindern, 0317 Oslo, Norway

Yoneyama, R.
Cardiovascular Research Center, Massachusetts General Hospital,
149 13th Street, Room 4215, Charlestown, MA 02129, USA

Yusa, K.
Department of Social and Environmental Medicine,
Graduate School of Medicine, Osaka University,
2-2 Yamadaoka, Suita, Osaka 565-0871, Japan

Wakayama, T.
Riken Center for Developmental Biology,
2-2-3 Minatojima-minamimachi, Kobe, 650-0047, Japan
(e-mail: teru@cdb.riken.jp)

Wang, S.
Division of Cell and Gene Therapy,
Departments of Neurology and Neurosurgery,
University of Rochester Medical Center,
601 Elmwood Avenue, Box 645, Rochester, NY 14642, USA

Washco, V.M.
Division of Cell and Gene Therapy,
Departments of Neurology and Neurosurgery,
University of Rochester Medical Center,
601 Elmwood Avenue, Box 645, Rochester, NY 14642, USA

Windrem, S.
Division of Cell and Gene Therapy,
Departments of Neurology and Neurosurgery,
University of Rochester Medical Center,
601 Elmwood Avenue, Box 645, Rochester, NY 14642, USA

Wu, Q.
Department of BioScience, Tokyo University of Agriculture,
Setagaya-ku, Tokyo 156-8502, Japan

Zenke, M.
Institute for Biomedical Engineering, Department of Cell Biology,
Aachen University Medical School,
Pauwelsstrasse 20, 52074 Aachen, Germany
(e-mail: martin.zenke@rwth-aachen.de)

1 What Is a Stem Cell Niche?

S.-I. Nishikawa, M. Osawa

1.1 The Niche for Quiescent Melanocyte Stem Cells 2
1.2 Advantage of the Melanocyte for Studying Niche 3
1.3 When Are Melanocyte Stem Cells Generated? 5
1.4 Differentiation of Melanocyte Stem Cells
    to Their Immediate Progenies Is Reversible 7
1.5 Characteristics of Quiescent Melanocyte Stem Cells 7
1.6 Low Level of Housekeeping Gene Expression
    in Quiescent Stem Cells 8
1.7 Wnt Signal Inhibition Is Implicated as a Mechanism
    Underlying Downregulation of Genes That Are Essential
    for Melanocyte Development 9
1.8 High Notch Expression in Quiescent Stem Cells 11
1.9 Current Model and Niche 11
References 13

Abstract. Niche has become the most important issue in stem cell biology, but it is still a hypothetical notion that cannot be defined in a better way than the microenvironment surrounding stem cells. Using a melanocyte stem cell system as a model, we have analyzed the cellular and molecular requirements for differentiation of quiescent stem cells. Our results demonstrate the multiple subsets within the stem cell compartment and thus suggests the complexity of niche.

1.1 The Niche for Quiescent Melanocyte Stem Cells

"Niche" is now becoming an important term in stem cell biology, describing the microenvironment that is specialized for maintaining the characteristic features of stem cells such as self-renewal or quiescence. This term has gained popularity from the early phase of its use, as it is a convenient term depicting an invariant concept of stem–microenvironmental interactions. For example, a recent review by Suda and his colleagues depicted the niche for hematopoietic stem cells (HPSCs) as pits, each of which fits to accommodate single stem cells (Fig. 1) (Suda et al. 2005). Indeed, such simplicity might be an important factor for any concept to be gain popularity. For instance, this scheme is able to articulate the mechanisms by which to limit the number of stem cells (SCs), which fits well to such a system as gut epithelium where only a few SCs are present in each villus (Potten et al. 2003). Even in this simple system where the histological organization is well characterized, the nature of the microenvironment, be it cells and/or matrix, is totally obscure. More difficult is to investigate the niche in a system such as a hematopoietic system in which histological localization of SCs has

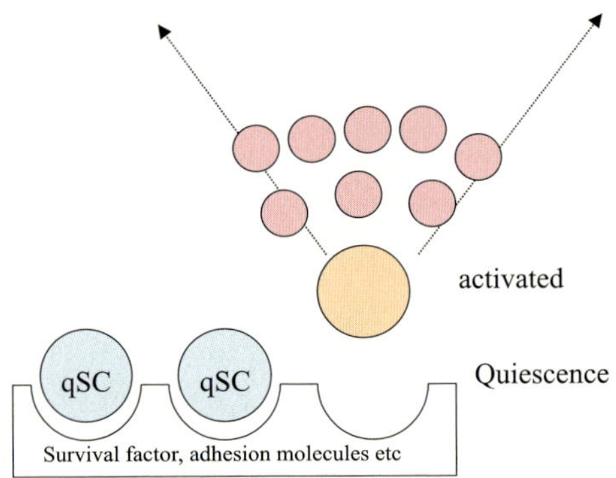

Fig. 1. A pit model for stem cell niche. The most popular model for stem cell niche. Niche is considered to be a microenvironment that supports quiescent SCs. When cells are released from this area, they become activated, proliferate, and undergo irreversible differentiation

been difficult to determine. Recent studies implicated osteoblast as the niche cell for hemapoietic cells (HPCs) (Arai et al. 2004; Calvi et al. 2003; Zhang et al. 2003). However, bone marrow is full of osteoblasts. If there is difference among the osteoblasts in terms of the ability to serve as a niche, what is the mechanism underlying generating a diversity of osteoblast populations? In the gut epithelium, it has been assumed that progenies derived directly from stem cells, thus locating adjacent to the stem cell, can contribute to direct which cells become stem cells by lateral inhibition mechanisms. In this situation, the niche is more self-assembling than what is prepared independently. In contrast, in HPCs, stem cells and their progenies may not form an integrated tissue, so that involvement of progenies in directing HPSCs is unlikely. If so, then, what directs osteoblasts to acquire the niche activity for HPSCs? It could be that the HPSC itself directs osteoblasts to be the niche for HPSCs. If, however, HPSC is required for niche formation, then how can the niche limit the number of HPSCs? Or it may be that there is no limit to the number of stem cells in the hematopoietic system. As such, the stem cell system may not be as simple as being able to be expressed by a simple pit-like structure. Thus, in this review, we attempt to decompose the process of stem cell differentiation by taking the melanocyte (MC) as a model, as it will tell us the molecular mechanisms required for the niche formation.

1.2 Advantage of the Melanocyte for Studying Niche

Two stem cell systems operate in a hair follicle: one is a stem cell system that gives rise to keratinocytes and hair appendage cells, the follicular stem (FS) and the other is the one that gives rise to melanocytes, the melanocyte stem (MCS). The hair follicle is unique in that it period-ically repeats the regeneration cycle throughout life and the stem cell system is segregated from proliferating compartments. Hair follicles are divided into two parts: one is the permanent portion whose architecture is maintained over the repeated hair regeneration cycles and the other is the transient portion that is lost and regenerates anew in each regener-ation cycle (Fig. 2). It was shown that the most immature SCs in both systems stay quiescent before a new regeneration cycle is initiated. In this sense, two SC systems are the most typical examples in which SC remains quiescent. Previous studies demonstrated that both the FS and MCS are located at the lower region of the permanent portion of the hair follicle (LPP) that has been called the bulge region (Alonso and Fuchs 2003; Gambardella and Barrandon 2003; Nishimura et al. 2002). In contrast, cells that are actively proliferating locate in the hair matrix at the bottom of the transient portion of the hair follicle. It is known that proliferation and differentiation of cells in the hair matrix is regulated by the mesenchymal cells present in the hair papillae (Jahoda et al. 1984; Kishimoto et al. 2000). Just as it is known to be the source of stem cell factor (SCF) that is essential for MC survival, it is likely that hair papillae are also the direct regulators of MC activity in the hair matrix. Along with morphogenetic elongation of the transient portion into sub-

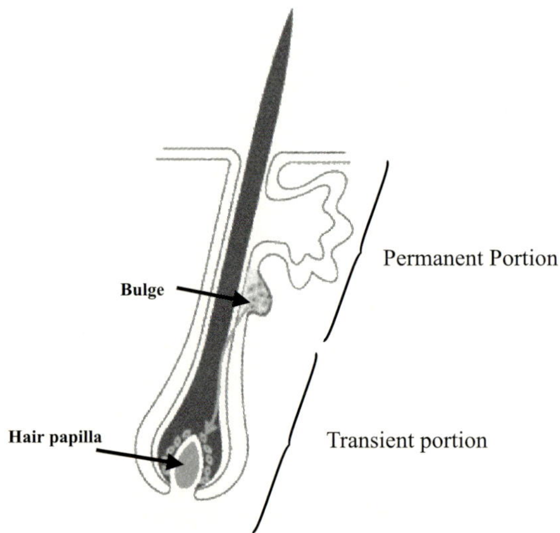

Fig. 2. Tissue organization of a hair follicle. A hair follicle is divided into two parts. The permanent portion that is the upper part of the bulge region is maintained throughout life, whereas the transient portion undergoes apoptosis and a regeneration cycle for each hair cycle. Hair papillae that consist of mesenchymal cells are essential for driving this hair regeneration cycle

cutaneous tissue, the hair papillae supporting active proliferation of the hair bulb become separated from the LPP.

In addition, MCS is unique in that it is proven to be able to exist alone without interacting with the MC lineage. Indeed, by the treatment to reduce the number of neonatal MCs, it is possible to observe hair follicles containing only a single MCSC in LPP. In this situation, if MCSC requires its niche, it is provided only by the keratinocytes and matrix. On the other hand, FS is surrounded by the keratinocytes, which are derived from FSs themselves. Thus, it is possible that the cellular organization of the niche for FSs is more complex. As such, the MC stem cell system can be regarded as a system in which the architecture of the stem cell niche would be the simplest among many stem cell systems in terms of tissue organization.

1.3 When Are Melanocyte Stem Cells Generated?

MCs are derived from neural crest cells, spread over the entire body surface through dermal mesenchyme, migrate into developing epidermis, are incorporated into developing hair follicles, and eventually undergo differentiation to MCSCs in LPP. As we have described this embryonic process in detail elsewhere (Yoshida et al. 1996), we will focus only on the neonatal processes of MCSC differentiation from embryonic progenitors. Of note is that in each hair follicle, MC progenitors derived from embryonic development give rise to both differentiated MCs and MCSCs. This means that mature MCs delivering pigment to the first hair of neonates are derived directly from embryonic progenitors rather than from MCSCs. Indeed, the mode that the stem cell comes latest appears to be common to most stem cell systems. For instance, while the HPCs appear from an early stage of embryogenesis, HPSCs are known to be generated later.

After knowing the location of MCSCs, the first question to be addressed is when the first MCSC is differentiated. (While whisker follicles can be the first site of MCSC differentiation, we will not consider this tissue in this article). In our previous study, we defined MCSs as cells that can survive in the absence of c-Kit signal. When an antagonistic mAb to c-Kit is injected immediately after birth, all first hair becomes depigmented, indicating that recruitment of mature MCs to hair follicles proceeds during neonatal days, and this process is c-Kit-dependent (Nishimura et al. 2002). Histologically, no melanocytes, be they mature or immature, are detectable in all hair bulbs. In the LPP of the guard hair that constitutes 20% of dorsal hairs, a few MCs that are marked by TRP2 expression are detectable. Indeed, these remaining MCs are MCSs, because pigmented hairs are regenerated from such guard hairs, whereas all other hair follicles remain depigmented throughout life. From these data, we proposed that differentiation of MCSCs in the guard hair is completed at the time of birth.

Recently, however, evidence indicating that differentiation of MCSs may complete a couple of days later came from the analysis of $Bcl2^{-/-}$ mouse. $Bcl2^{-/-}$ mice born with pigmented hair but pigmentation is lost from most hair after the first hair regeneration cycle. A recent study conducted by Nishimura et al. demonstrated that MC lineages are found in LPP up to day 4 after birth, whereas they are lost already at P6 (Nishimura et al. 2005). This observation was interpreted to mean that Bcl2 is required for maintenance of nascent MCSCs in the LPP. On the contrary, we recently reported that neonatal blockade of c-Kit function resulted in complete depletion of functional MCSs, though we also confirmed the presence of MC in LPP of the $Bcl2^{-/-}$ mouse. Moreover, we also showed that MCSs that give rise to mature MCs in the next hair cycle are completely rescued if expression of SCF, the ligand for c-Kit, is maintained after birth by expression of the SCF transgene in the dermal keratinocytes, which otherwise is downregulated during neonatal days (Mak et al., unpublished observation). From these data, we concluded that Bcl2 is required for short-term survival of the MC stage when SCF is rapidly downregulated from the epidermis, whereas it is not required after MCSC differentiation is completed. All these observations indicate that, while exact timing of the MCS differentiation is yet to be determined, it may be completed sometime at day 3–4 after birth. In such a situation as the $Bcl2^{-/-}$ mouse, this time interval becomes clear as most MCs cannot survive over this SCF-negative stage.

1.4 Differentiation of Melanocyte Stem Cells to Their Immediate Progenies Is Reversible

In the hierarchical architecture of the stem cell system, the exit of SCs from the quiescent stage to the proliferating stage has been assumed to be irreversible process. However, recent studies suggested that in many stem cell systems, at least some of the immediate progenies of SCs preserve a potential to give rise to new SCs (Marshman et al. 2002). In the MC system, we have also shown that MCSCs and their immediate progenies are reversible.

This was shown using a following complicated experimental system, with the details published in Nishimura et al. (2002). In adult mouse, all cells of MC lineage are present only in the hair follicles. This is because SCF, which is essential for survival and differentiation of MCs, is downregulated from all keratinocytes in the upper part of the permanent portion including interfollicular epidermis. In fact, MCs in these regions are restored in the transgenic mouse that express SCF constitutively in all keratinocytes. Thus, in this transgenic mouse, MCSCs are recruiting MCs to both the hair matrix and interfollicular epidermis. When c-Kit function is blocked at the neonatal stage of this transgenic mouse, we can create a situation in which only guard hairs contain MCSs while other types do not. Interestingly, in this situation, MCs derived from MCSCs do migrate upward, colonize to the surrounding hair follicles where MCs are completely depleted, and eventually form a new MCSC system. This demonstrates that at least a portion of MCs that have exited from the quiescent SC stage (qSC) can return to qSC when they meet an appropriate microenvironment in the new follicles. This finding also supports the notion that MCS differentiation is directed extrinsically by the microenvironment, which is formed in the LPP independently from the MC lineage.

1.5 Characteristics of Quiescent Melanocyte Stem Cells

Based on the above-mentioned observations, we proposed a three-compartment model of the MC stem cell system, which consists of qSC, aMC that are not quiescence but still preserve an ability to become qSC, and differentiated cells (DCs) that have undergone irreversible commitment to differentiated MCs.

With this scheme in mind, we have attempted to define the difference between qSCs, aMCs, and DCs by analyzing the gene expression profile of each stage. The detail of our method to analyze the gene expression profile of single cells representing each compartment has been described elsewhere (Osawa et al. 2005). A summary of the results derived from this study follows:

1. qSCs are characterized by its low expression of a set of housekeeping genes such as β-actin and GPAH.
2. Genes essential for melanocyte development such as Sox10, Mitf, TRP, and c-Kit are downregulated upon differentiation to qSC, while Pax 3 are constantly expressed.
3. A set of Wnt inhibitors are expressed higher than other compartments.
4. Notch and Hes are expressed higher in qSC than other compartments.

In this article, we will present current findings of our research in these four directions. However, we have also noted other specific qSC features that are potentially interesting but remain largely for future research.

1.6 Low Level of Housekeeping Gene Expression in Quiescent Stem Cells

The main feature of qSCs, which was noticed first by comparing gene expression profiles between qSC and other compartments was a low level expression of housekeeping genes such as β-actin, γ-actin, Gapdh, and Aldoa. By using the single-cell PCR method, however, it is difficult to avoid variation among assays. Hence, we attempted to evaluate this possibility by using a strain of mouse in which MCs express green fluorescent protein (GFP) driven by chicken actin promoter. If qSCs express a lower level of β-actin, qSCs can be distinguished from other compartments by using the GFP level as an indicator. As expected, we could detect two populations that can be clearly separated by FACS. Indeed, the expression levels of the two populations differ five to ten-fold, and further analysis of the sorted populations from expression analysis of other markers is consistent with the notion that GFP[low] and GFP[high] cells represent qSCs and other compartments, respectively. Interestingly, relatively large variation was observed both in GFP^{low} and GFP^{high} populations. This is consistent with our results on single-cell PCR, suggesting the presence of variations in the expression level of housekeeping genes in each compartment.

The CAG promoter used in this study is a combination of CMV and CAG promoters, which contains six SP1, two NFB, one C/EBP, and two CREB sites. Hence, the low level of housekeeping genes in qSCs is likely to be ascribed to the complex of transcriptional regulators assembled in this region. Of particular interest in this sense is that expression of IκB and Cri1, which inhibit the activity of NFκB and CREB, respectively, are expressed at a high level. We are currently exploring the possibility that these molecules are involved in the suppression of housekeeping gene expression. In addition, we are currently exploring whether or not this feature of housekeeping gene expression is also common to other stem cell systems. If this is the case for other stem cell systems, the expression level of housekeeping genes will be a useful marker for distinguishing qSCs.

1.7 Wnt Signal Inhibition Is Implicated as a Mechanism Underlying Downregulation of Genes That Are Essential for Melanocyte Development

It was shown that expression of Pax3 and Sox10, which are induced upon melanocyte commitment, induce Mitf that triggers a set of genes that are essential for melanocyte differentiation. However, we found that most genes that are triggered by Mitf are downregulated in qSCs. Moreover, expression of Mitf and Sox10 by themselves are also downregulated. Hence, it is likely that many molecules characterizing MC lineage are turned off at the qSC stage. In contrast, we detected Pax3 expression at the comparable level with other compartments. This is consistent with a recent study by Epstein and his colleagues suggesting that Pax3 is essential for maintenance of the immature stage as it blocks Mitf, which is the major inducer of MC differentiation (Lang et al. 2005). According to this model, Pax3, while it is the essential molecule for induction of Mitf, also inhibits Mitf activity by competing the same region where the binding site for Mitf and Sox10 are present. Furthermore, they proposed that this situation can be reversed by the Wnt signal, as binding of activated β-catenin to Lef/TCF in this site displaces Pax3 and recruits Mitf and Sox10. Thus, in this scheme, the Wnt signal is the switcher of the Pax3-directed immature state and Mitf-directed differentiated state. Our finding that genes that are downstream of Mitf are expressed at a low level is consistent with the notion that Mitf activity is inhibited in the SC compartment. However, it should be noted that qSC by our definition is different from that defined by Epstein et al. in that Sox10 and Mitf expression themselves are downregulated at the transcriptional level in qSCs rather than inactivated. Thus, while the Wnt signal is involved in the decision making toward terminal differentiation of MCs, it may not have a role in differentiation into qSCs. We therefore propose a three compartment model (Fig. 3) instead of the two-compartment model of Epstein's group. While the biological significance of low Sox10 and Mitf transcription in the qSC compartment remains to be determined, we consider that this process would be necessary for final differentiation to the quiescent state. In conclusion, we define qSCs as those that express a low level of Sox10 and Mitf. After exiting from the quiescent state,

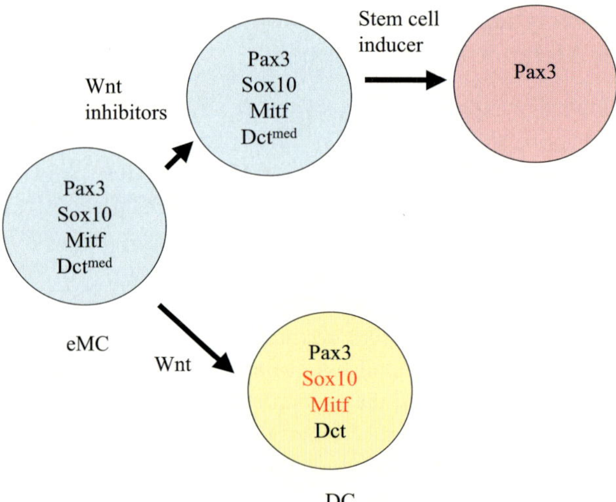

Fig. 3. Three-compartment model for melanocyte SCs. For details, see text

those molecules, essential for MC differentiation, are re-expressed and their activity comes under the control of the Wnt signal.

It should be noted, however, that Wnt may not be essential for embryonic MC development. For instance, according to the two-state model of Epstein et al., Dct expression, an indicator of Mitf activity, is not expressed in MCs that remain at the immature state due to lack of a Wnt signal. However, during embryogenesis, most if not all MCs express Dct. In fact, Saito et al. demonstrated that Mitf can interact with LEF1 to induce Dct, not only in the presence of β-catenin, but also in its absence (Saito et al. 2003). Hence, it is likely that Dct is transcribed in the absence of the Wnt signal. Hence, we speculate that the Wnt signal may play a role in the final commitment toward irreversible differentiation. In fact, hair papilla is implicated as the major source of the Wnt signal in the hair follicle, but epidermal keratinocytes have not been shown to express Wnt. If so, this suggests that a strong Wnt signal is available only after the initiation of hair follicle formation, thus the model of Epstein et al. is valid for the process in the hair matrix.

1.8 High Notch Expression in Quiescent Stem Cells

While Notch was shown to be expressed in melanoma cells (Hoek et al. 2004), the Notch signal has not been implicated in MC development. Our analysis on the gene expression profile demonstrated that Notch is expressed at a higher level in qSC. We therefore investigated the role of Notch in MC development as well as the maintenance of qMC by using various strategies. As details of our study are described in Moriyama et al. (in press) paper, we want emphasize that our study demonstrated that the Notch signal is essential for embryonic MC development and may also be important for the maintenance of qMCs.

1.9 Current Model and Niche

In the final section, I would like to present our hypothesis on the process during MC differentiation of qSCs in the neonate (Fig. 3) and speculate on the molecular nature of niche.

Upon differentiation to MC, MCs are characterized by the expression of upstream molecules Pax3, Sox10, and Mitf. This combination by itself is sufficient to induce a set of downstream genes required for MC differentiation. While Wnt can modulate this system, it is likely that Wnt does not play a significant role in MC development in the embryonic skin. Indeed, we think that the source of Wnt may not present in the dermal region before hair follicle formation. Wnt is expressed in mesenchymal cells that induce hair follicles and will eventually develop to hair papillae. Hence, only at the time of hair follicle formation are MCs exposed to the Wnt signal that is emanated from hair papillae. Given that Wnt is the signal directing the irreversible differentiation, a high concentration of Wnt is deleterious for maintenance of immaturity. Thus, in order to maintain the immature state, Wnt concentration should be maintained at the level lower than that needed to induce irreversible differentiation. This might be attained by the distance from hair papillae, the Wnt source, but the role of active Wnt inhibitors is also implicated. In fact, it was reported that a set of Wnt inhibitors are expressed in the keratinocytes in the LPP. This inhibition of the Wnt signal may be required for protecting MCs from undergoing irreversible commitment to differentiation. Wnt inhibition, however, is not sufficient to induce the final step of qSC differentiation, as both Sox10 and Mitf are downregulated at the transcriptional level during this process. Moreover, the cell cycle of MC is arrested at G1 and eventually introduced to the quiescent state. Our gene expression study also showed that a set of housekeeping genes are downregulated during this process. While downregulation of SCF also occurs at the same time in the microenvironment of LPP, this downregulation may have nothing to do with the differentiation of qSCs, because qSC differentiation does occur in the transgenic mice that express SCF constitutively in dermal keratinocytes. Thus, quiescence may not be attained passively by blocking the availability of Wnt or SCF, but rather induced actively by yet unknown mechanisms. At present, it is difficult to determine the order of events during qSC differentiation: for instance which is earlier, acquisition of quiescence or downregulation of Mitf/Sox10? Nonetheless, it is clear that the differentiation of qSC can be decomposed into multiple processes, which are likely to be regulated by various molecules.

As discussed above, the microenvironment supporting qSC differentiation is required to support:

1. Inactivation of Mitf/Sox10 activity by Wnt inhibition
2. Downregulation of Sox10 and Mitf transcription
3. Downregulation of transcription of house-keeping genes
4. Introduction of quiescence
5. Maintenance of the survival of qSCs

Furthermore, it should not be inhibitory to the process of reactivation, even though active suppression mechanisms are required for the induction of quiescence. Whether all these functions are concentrated on a single cell type or mediated by multiple cells present in the same area is an issue to be explored in the future. Nonetheless, we could not elucidate this process without knowing the molecules responsible for each process.

References

Alonso L, Fuchs E (2003) Stem cells of the skin epithelium. Proc Natl Acad Sci U S A 100:11830–11835

Arai F, Hirao A, Ohmura M, Sato H, Matsuoka S, Takubo K, Ito K, Koh GY, Suda T (2004) Tie2/angiopoietin-1 signaling regulates hematopoietic stem cell quiescence in the bone marrow niche. Cell 118:149–161

Calvi LM, Adams GB, Weibrecht KW, Weber JM, Olson DP, Knight MC, Martin RP, Schipani E, Divieti P, Bringhurst FR et al. (2003) Osteoblastic cells regulate the haematopoietic stem cell niche. Nature 425:841–846

Gambardella L, Barrandon Y (2003) The multifaceted adult epidermal stem cell. Curr Opin Cell Biol 15:771–777

Hoek K, Rimm DL, Williams KR, Zhao H, Ariyan S, Lin A, Kluger HM, Berger AJ, Cheng E, Trombetta ES et al. (2004) Expression profiling reveals novel pathways in the transformation of melanocytes to melanomas. Cancer Res 64:5270–5282

Jahoda CA, Horne KA, Oliver RF (1984) Induction of hair growth by implantation of cultured dermal papilla cells. Nature 311:560–562

Kishimoto J, Burgeson RE, Morgan BA (2000) Wnt signaling maintains the hair-inducing activity of the dermal papilla. Genes Dev 14:1181–1185

Lang D, Lu MM, Huang L, Engleka KA, Zhang M, Chu EY, Lipner S, Skoultchi A, Millar SE, Epstein JA (2005) Pax3 functions at a nodal point in melanocyte stem cell differentiation. Nature 433:884–887

Marshman E, Booth C, Potten CS (2002) The intestinal epithelial stem cell. Bioessays 24:91–98

Moriyama M, Osawa M, Siu-Shan M et al. (2006) Notch signaling via Hes1 transcription factor maintains survival of melanoblasts and melanocyte stem cells. J Cell Biol, in press

Nishimura EK, Granter SR, Fisher DE (2005) Mechanisms of hair graying: incomplete melanocyte stem cell maintenance in the niche. Science 307:720–724

Nishimura EK, Jordan SA, Oshima H, Yoshida H, Osawa M, Moriyama M, Jackson IJ, Barrandon Y, Miyachi Y, Nishikawa S (2002) Dominant role of the niche in melanocyte stem-cell fate determination. Nature 416:854–860

Potten CS, Booth C, Tudor GL, Booth D, Brady G, Hurley P, Ashton G, Clarke R, Sakakibara S, Okano H (2003) Identification of a putative intestinal stem cell and early lineage marker; musashi-1. Differentiation 71:28–41

Saito H, Yasumoto K, Takeda K, Takahashi K, Yamamoto H, Shibahara S (2003) Microphthalmia-associated transcription factor in the Wnt signaling pathway. Pigment Cell Res 16:261–265

Suda T, Arai F, Hirao A (2005) Hematopoietic stem cells and their niche. Trends Immunol 26:426–433

Yoshida H, Kunisada T, Kusakabe M, Nishikawa S, Nishikawa SI (1996) Distinct stages of melanocyte differentiation revealed by analysis of nonuniform pigmentation patterns. Development 122:1207–1214

Zhang J, Niu C, Ye L, Huang H, He X, Tong WG, Ross J, Haug J, Johnson T, Feng JQ et al. (2003) Identification of the haematopoietic stem cell niche and control of the niche size. Nature 425:836–841

2 New Strategy for Comprehensive Analysis of Gene Functions in Embryonic Stem cells

J. Takeda, K. Horie, K. Yusa

2.1   Strategy to Introduce Bi-Allelic Mutations   16
2.2   Insertional Mutagenesis and Screening for Genes Involved
      in Stem Cell Maintenance in Embryonic Stem Cells   18
References   21

Abstract. At present, the limitation of Phenotype-based genetic screening in embryonic stem cells (ESCs) is the diploid nature of the genome. Since it is known that cells deficient in the Bloom's syndrome gene (*Blm*) show an increased rate of homologous recombination, we have developed a new system to conditionally regulate the *Blm* allele for introduction of bi-allelic mutations across the genome. Transient deficiency of *Blm* induces homologous recombination not only between sister chromatids but also between homologous chromosomes, resulting in a high rate of loss of heterozygosity (LOH). Introduction of genome-wide mutations in ESCs can be achieved by retroviral vector. In combination, using genome-wide mutagenesis and transient loss of *Blm* expression, we have generated ES libraries with bi-allelic mutations. These results show that this new system is very efficient for identifying gene functions in ESCs.

2.1 Strategy to Introduce Bi-Allelic Mutations

Loss of function screening in mouse or its cultured cell lines has been difficult due to the diploid nature of the genome. We have chosen embryonic stem cells (ESCs) because they have the potential to differentiate into many different cell types. In addition to this characteristic, ESCs most importantly maintain their normal karyotype, which is appropriate for bi-allelic mutagenesis (Guo et al. 2004). We are interested in the Bloom's syndrome phenotype, a rare genetic disorder caused by the mutation of *BLM*, which results in predisposition to cancer. This predisposition is associated with the increased rate in LOH. Generation of a conditional *Blm* allele would be important for the prevention of continuous accumulation of bi-allelic mutations and genomic instability (Yusa et al. 2004). In fact, two out of three *Blm*-null alleles generated caused embryonic lethality (Chester et al. 1998; Goss et al. 2002). Although mice bearing the third allele were viable (Luo et al. 2000), leakage of *Blm* expression has been recently reported (McDaniel et al. 2003).

The molecular mechanism of how BLM works during homologous recombination has been discussed. When double-stranded breaks occur, BLM together with topoisomerase IIIα suppresses crossing over during homologous recombination, thereby preventing LOH (Wu and Hickson 2003).

The mouse *BLM* homolog, *Blm*, was modified by the insertion of a tetracycline cassette (tet cassette) in order to regulate *Blm* in a reversible manner; the schematic diagram in Fig. 1 shows the modified *Blm* containing the tet cassette. The tet cassettes were inserted immediately upstream of the translational initiation sites of both alleles of *Blm* to generate *Blm*[tet]. Regulation of *Blm* expression was examined with use of a potent tetracycline analog, doxycycline (dox). Addition of dox in ESCs bearing *Blm*[tet] resulted in immediate reduction of *Blm*. In the absence of Blm, elevated sister chromatid exchange (SCE), a hallmark of BLM deficiency, was observed (Fig. 2). Lack of BLM also leads to an elevated rate of crossing over. To test the effect of transient loss of Blm on the cross over rate, we inserted a mutant *Neo* gene into the *Fas ligand* locus in *Blm*[tet] cells. High dosage of G418 ($\sim$ 1 mg/ml) selects for only cells bearing bi-allelic mutant *Neo* genes but not those bearing the mono-allelic gene (Koike et al. 2002). The rate of crossing-over was

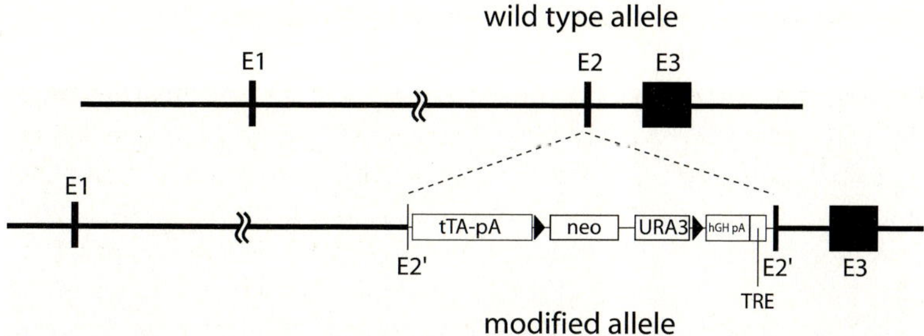

Fig. 1. Generation of modified Blm allele. A tet cassette containing the Neo gene was inserted upstream of the translational initiation site of Blm to generate a modified Blm allele

significantly higher (approximately 30-fold) than in control cells (Yusa et al. 2004).

To demonstrate the practical applicability of this technology, an ESC library containing bi-allelic mutations throughout the genome was constructed using N-ethyl-N-nitrosourea (ENU) as a mutagen. We screened the ESC library for mutants deficient in glycosylphosphatidylinositol (GPI) -anchor biosynthesis. Genes involved in GPI-anchor biosynthesis are widely distributed in the genome and mutant cells are positively selected by the use of aerolysin, which kills wild-type cells with GPI-

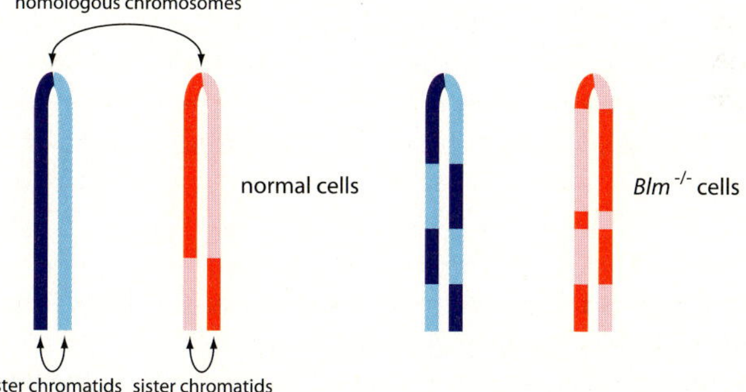

Fig. 2. Schematic presentation of elevated SCE in Blm-deficient cells. At the 4N stage, elevated sister chromatid exchange (SCE) is observed in Blm-deficient cells, but not in normal cells

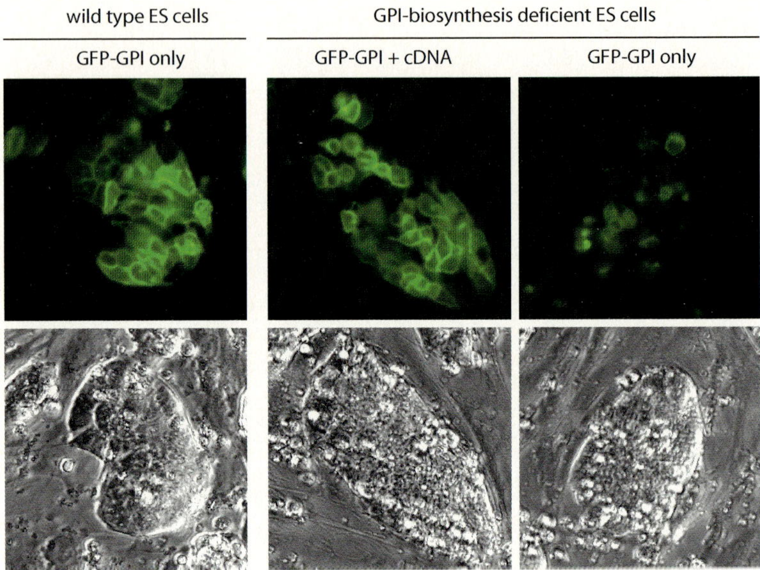

Fig. 3. Complementation analysis of GPI-anchor-deficient cells to identify a defective cDNA. GPI-anchored GFP proteins were expressed on the cell surface only when a defective cDNA was provided. *Left panel* shows positive control

anchor. Thirty-five GPI-anchor-deficient mutants were isolated (Yusa et al. 2004). To classify these mutants, the GPI-anchored GFP gene was co-transfected with complementary cDNA involved in GPI-anchor biosynthesis (Fig. 3). Identified mutant genes were widely distributed throughout the genome as shown in Table 1 and more than half of known genes were obtained (Yusa et al. 2004), indicating the high efficiency of this system.

2.2 Insertional Mutagenesis and Screening for Genes Involved in Stem Cell Maintenance in Embryonic Stem Cells

To identify novel genes responsible for stem cell maintenance in ESCs, we constructed a gene-trap retrovirus vector as shown in Fig. 4. The

Table 1. Chromosome number of mutated genes

Gene	Chromosome
PigC	1
PigM	1
PigU	2
DPM1	2
GPI8	3
PigO	4
PigV	4
Unknown 1	7
PigB	9
PigS	11
PigH	12
GAA1	15
Unknown 2	16
PigA	X

splice acceptor (*SA*) and internal ribosome entry site *(IRES)* followed by the *Neo* gene was inserted in reverse orientation to the retrovirus vector backbone. The integrated retrovirus vector contains two *FRT* sites in both of the long terminal repeats, allowing for the reversal of mutant phenotype by excision with Flp recombinase. To investigate the molecular mechanism(s) of how ESCs are maintained as stem cells, we inserted a cassette consisting of puromycin-resistant gene (*Puro*) followed by the *IRES GFP* gene into the *Nanog* locus in *Blm*^{tet} ESCs (*Puro-GFP-Nanog* ESC). Nanog is a critical factor for maintaining pluripotency of ESCs independently of the LIF/Stat3 pathway and its expression is restricted in ESCs (Chambers et al. 2003; Mitsui et al. 2003). As long as ESCs are maintained as stem cells, they should be puromycin-resistant. ESCs mutagenized with retrovirus vector followed by introduction of bi-allelic mutations with dox can be screened for pluripotency in the absence of LIF. In this situation, activation of the Nanog pathway is absolutely required.

The Wnt signal pathway is known to affect the differentiating potential of ESCs. The adenomatous polyposis coli gene (*Apc*) is a major component of the Wnt pathway and inhibits the differentiation of ESCs (Kielman et al. 2002). We therefore disrupted both alleles of *Apc* genes in *Puro-GFP-Nanog* ESCs (*Apc*-ESCs) and used it as a positive

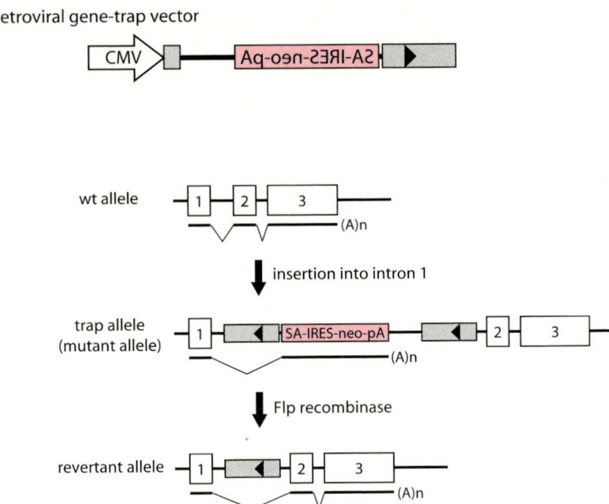

Fig. 4. Reversible gene-trap mutagenesis. If a retroviral vector is inserted into the first intron of a gene, the gene would be disrupted to create a mutant allele. Flp recombinase removes the trap unit in the vector to create the revertant allele

control for stem cell maintenance without LIF. A protocol for the selection of pluripotency without LIF is shown in Fig. 5. As a control experiment, *Apc*-ESCs were mixed with *Puro-GFP-Nanog* ESCs (ratio, $1 : 10^5$) and the mixture was selected according to the protocol. *Apc*-ESCs were highly enriched after selection, suggesting that the protocol is appropriate for obtaining pluripotent ES clones without LIF. Dox-treated *Puro-GFP-Nanog* ESCs infected with gene-trap retrovirus were selected according to the protocol and more than 600 clones were collected. Chromosomal insertion sites were determined by ligation-mediated PCR (LM-PCR). Five genes had multiple hits, ranging two to seven hits per gene. Currently, we are intensively analyzing these genes.

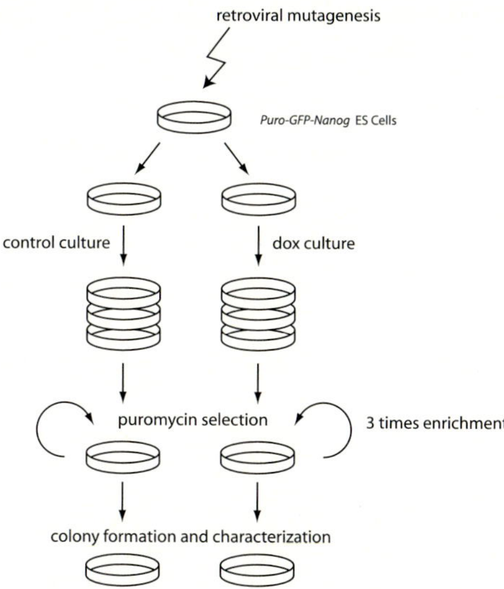

Fig. 5. Screening of pluripotent ESCs in the absence of LIF from an ESC library bearing bi-allelic mutations

References

Chambers I, Colby D, Robertson M, Nichols J, Lee S, Tweedie S, Smith A (2003) Functional expression cloning of Nanog, a pluripotency sustaining factor in embryonic stem cells. Cell 113:643–655

Chester N, Kuo F, Kozak C, O'Hara CD, Leder P (1998) Stage-specific apoptosis, developmental delay, embryonic lethality in mice homozygous for a targeted disruption in the murine Bloom's syndrome gene. Genes Dev 12:3382–3393

Goss KH, Risinger MA, Kordich JJ, Sanz MM, Straughen JE, Slovek LE, Capobianco AJ, German J, Boivin GP, Groden J (2002) Enhanced tumor formation in mice heterozygous for Blm mutation. Science 297:2051–2053

Guo G, Wang W, Bradley A (2004) Mismatch repair genes identified using genetic screens in Blm-deficient embryonic stem cells. Nature 429:891–895

Kielman MF, Rindapaa M, Gaspar C, van Poppel N, Breukel C, van Leeuwen S, Taketo MM, Roberts S, Smits R, Fodde R (2002) Apc modulates embryonic stem-cell differentiation by controlling the dosage of beta-catenin signaling. Nat Genet 32:594–605

Koike H, Horie K, Fukuyama H, Kondoh G, Nagata S, Takeda J (2002) Efficient biallelic mutagenesis with Cre/loxP-mediated inter-chromosomal recombination. EMBO Rep 3:433–437

Luo G, Santoro IM, McDaniel LD, Nishijima I, Mills M, Youssoufian H, Vogel H, Schultz RA, Bradley A (2000) Cancer predisposition caused by elevated mitotic recombination in Bloom mice. Nat Genet 26:424–429

McDaniel LD, Chester N, Watson M, Borowsky AD, Leder P, Schultz RA (2003) Chromosome instability and tumor predisposition inversely correlate with BLM protein levels. DNARepair (Amst) 2:1387–1404

Mitsui K, Tokuzawa Y, Itoh H, Segawa K, Murakami M, Takahashi K, Maruyama M, Maeda M, Yamanaka S (2003) The homeoprotein Nanog is required for maintenance of pluripotency in mouse epiblast and ES cells. Cell 113:631–642

Wu L, Hickson ID (2003) The Bloom's syndrome helicase suppresses crossing over during homologous recombination. Nature 426:870–874

Yusa K, Horie K, Kondoh G, Kouno M, Maeda Y, Kinoshita T, Takeda J (2004) Genome-wide phenotype analysis in ES cells by regulated disruption of Bloom's syndrome gene. Nature 429:896–899

3 Paternal Dual Barrier by Ifg2-H19 and Dlk1-Gtl2 to Parthenogenesis in Mice

T. Kono, M. Kawahara, Q. Wu, H. Hiura, Y. Obata

3.1   Introduction 24
3.2   Analysis of Gametic Imprinting During Oocyte Growth 25
3.3   Role of Paternally Imprinted Genes on Parthenogenesis 26
3.4   Mechanism Responsible for Extended Development 27
3.5   Methylation Analysis of the IG-DMR at Dlk1-Gtl2 Domain 29
3.6   Conclusion 30
References 30

Abstract. The functional difference between the maternal and paternal genome, which is characterized by epigenetic modifications during gametogenesis, that is genomic imprinting, prevents mammalian embryos from parthenogenesis. Genomic imprinting leads to nonequivalent expression of imprinted genes from the maternal and paternal alleles. However, our research showed that alteration of maternal imprinting by oocyte reconstruction using nongrowing oocytes together with deletion of the H19 gene, provides appropriate expression of maternally imprinted genes. Here we discuss that further alteration of paternally imprinted gene expressions at chromosomes 7 and 12 allows the ng/fg parthenogenetic embryos to develop to term, suggesting that the paternal contribution is obligatory for the descendant.

3.1   Introduction

Parthenogenesis is not a peculiar reproductive strategy, not even among higher organisms. However, only mammals have relinquished parthenogenesis as a means of producing descendants solely from maternal germ cells. Consequently, parthenogenetic development, in which embryos contain exclusively maternal genomes, results in death at early gestation. Attempts to induce parthenogenetic development have consistently resulted in embryonic death in up to 10 days of gestation in mice (Fig. 1) (Surani and Barton 1983; McGrath and Solter 1984; Barton et al. 1984; Surani et al. 1990) and around 21 days of gestation in sheep (Hagemann et al. 1998) and pigs (Kure-bayashi et al. 2000). The failure of parthenogenetic development in these cases is a result of genomic imprinting, which is a unique molecular mechanism for controlling parental origin-specific gene expression, depending on DNA methylation status, in the differentially methylated region (DMR) for each imprinted gene (Feil et al. 1994; Sasaki et al. 1995; Repoche et al. 1997; Thorvaldsen et al. 1998; Hark et al. 2000). However, until recently there was some doubt whether genomic imprinting is only a critical barrier to parthenogenesis in mammals. Over the past decade, a series of our studies has reached this conclusion (Kono et al. 1996, 2002, 2004). First we have shown that embryos containing genomes from nongrowing (ng) and fully grown (fg) oocytes, i.e., ng^{wt}/fg^{wt} PE (wt, wild type), developed to E13.5 (Kono et al. 1996) (Fig. 1), suggesting that ng oocyte genome is naïve in that it has not been subjected to epigenetic modification during oocyte growth. This extended development was achieved by switching maternal genomic imprinting status to those of paternal status, by which paternally expressed genes are expressed from ng alleles. However, genes regulated by paternally imprinting during spermatogenesis are not modified in the ng^{wt}/fg^{wt} PE (Obata and Kono 1998). From these results, it was suggested that the parthenogenetic development could be extended to term by further regulation of paternally imprinted genes. Recently we obtained a live pup from ng^{wt}/fg^{wt} PE that harbored a deletion of the *H*19 transcription unit together with its differentially methylated region (DMR), in which paternally imprinted *Igf2* and *H*19 were expressed only from ng and fg alleles, respectively (Kono et al. 2004). Here, we discuss

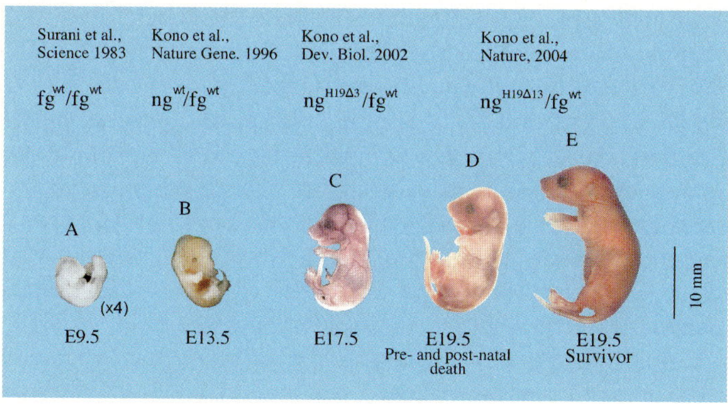

Fig. 1. Developmental outcome of reconstructed parthenogenetic embryos

how the parthenogenetic embryos extended their development to term although only a few of them survived and grew as a normal individual.

3.2 Analysis of Gametic Imprinting During Oocyte Growth

Analyses on DNA methylation have showed that the parent-of-origin-specific epigenetic modification of the chromatin is independently imposed during either spermatogenesis or oogenesis, following an erasure process during the differentiation of primordial germ cells (Kono et al. 1996; Mann 2001; Lucifero et al. 2004; Hiura et al. in press). This is supported by results from functional analysis of development and imprinted gene expressions using ng/fg PE (Kono et al. 1996; Bao et al. 2000; Obata and Kono 2002). The production of the embryos with maternal alleles that have not yet been subjected to maternal epigenetic modifications during oocyte growth provides the first viable opportunity to investigate the role of these modifications in embryonic development. Assessment of developmental ability showed that the ng^{wt}/fg^{wt} PE developed to E13.5 accompanyed with formation of functional placenta (Kono et al. 1996). Gene expression analysis by RT-PCR revealed that the extended development of ng/fg parthenotes was caused by the appropriate expression of maternally imprinted genes (Obata et al. 1998). We also investigated when and how, precisely, maternal primary imprinting is established using parthenogenetic embryos containing one genome from a nongrowing or growth stage oocyte from 1- to 20-day-old mice and one from a fully grown oocyte of adult mice (Obata and Kono 2002). Gene expression analysis of eight imprinted genes, *Peg1/Mest, Peg3, Snrpn, Znf127, Ndn, Impact, Igf2r,* and p57^{KIP2}, was conducted and showed that the epigenetic signals for each imprinted gene are not imposed all together at a specific time during oocyte growth, but rather occur within a wide range during the period from primary to antral follicle stage oocytes. Moreover, ng oocyte-derived X chromosome is preferentially inactivated in the extraembryonic tissues of the ng/fg PE because maternal primary imprinting during oocyte growth serves to render maternal X chromosome resistant to inactivation in the extraembryonic tissues (Tada et al. 2002). These results correspond well with our previous report demonstrating that fertilized embryos containing maternal genome from a growth stage oocyte extended their developmental competence gradually (Bao et al. 2000). Taking these data together, it was concluded that maternal imprinting is imposed during the oocyte growth period.

3.3   Role of Paternally Imprinted Genes on Parthenogenesis

A question that has arisen from a series of experiments suggested that the means by which an embryo is released from paternal genomic imprinting, which in turn leads to the normal development of parthenogenetic embryos containing only maternal genomes. To address the question of whether the collective expression of *Igf2* accompanied by *H*19 would result in the extended development of parthenotes, we used mice with deletion (H1913) of the *H*19 transcription unit together with its DMR (differentially methylated region) (Leighton et al. 1995) as an ng oocyte donor for the reconstruction of parthenogenetic embryos. In order to assess the extent of development, 343 morula/blastocysts derived from ng$^{H1913-KO}$/fg^{wt} oocytes were transferred to 26 recipient females, and autopsies were carried out to assess term development at 19.5 days of gestation (Kono et al. 2004). Beyond our expectations, a total of ten live and 18 dead pups were recovered (Fig. 1, Table 1). Surprisingly, of the ten live pups, two individuals were apparently morphologically normal

Table 1. Development of the ng/fg reconstructed parthenogenetic embryos

Genotypes	% Blastocysts	% E9.5	% E12.5	% E15.5	% E19.5
fg^{wt}/fg^{wt}	92	6	0	–	–
ng^{wt}/fg^{wt}	87	41	27	0	–
ng^{H193}/fg^{wt}	85	–	25	6	0
ng^{H1913}/fg^{wt}	92	45	23	19	12

and successfully restored, i.e., these two pups exhibited the apparently normal morphology of neonates. The body weights of the surviving parthenogenetic pups were 1,372 mg and 1,310 mg, which was similar to that of control $BDF1xB6^{H1913-KO}$ pups (1,326 mg). The other fetuses showed severe growth retardation and no apparent differences in phenotype, and the living fetuses were dead within 15 min. The body weight did not differ between the dead (786 mg, $n = 18$) and the live (744 mg, $n = 8$) pups. One of the two survivors was nursed by a foster mother, and showed normal growth, maturation, and, after mating, conceived and delivered normal pups. The other was used for gene expression analysis on the day of recovery.

In this type of parthenotes, the *Igf2* and *H19* genes were expressed and repressed, respectively, from the ng^{H1913} alleles (Wu et al. 2006). These embryos successfully developed to term; however, with the exception of two pups, all parthenogenetic pups showed severe growth retardation and died before or shortly after their recovery from the uterus at 19.5 days of gestation.

3.4   Mechanism Responsible for Extended Development

Among the issues raised by these results, of primary importance is the elucidation of how and why these $ng^{H1913-KO}/fg^{wt}$ parthenotes were able to develop and survive to adulthood (Kono et al. 2004). In this regard, the data from our global gene expression analysis by oligo microarray were quite intriguing. The silencing and activation of *H19* and *Igf2*, respectively, in the ng alleles led to the rectification of expression patterns of a wide range of genes, including other imprinted genes in the $ng^{H1913-KO}/fg^{wt}$ parthenotes. A quantitative expression analysis by real-time RT-PCR using major organs showed that the expression patterns of *Igf2* and *H*19 obtained were similar to those of the controls except the low expression of the *Igf2* in the liver (Wu et al. 2006). In all organs of the ng^{H1913}/fg^{wt} PE, *Gtl2* was overexpressed to approximately twice the control level, showing a peak at E15.5. However, *Dlk1* expression was repressed in brain, tongue, heart, liver, and leg, with expression at levels below 20% of the controls. With regard to the tested organs, the expression levels of the paternally expressed *Dio3* in the ng^{H1913}/fg^{wt} PE were approximately 70% of the controls, with the exception of the tongue (<50%), while the maternally expressed *Mirg* was overexpressed in most tissues of the ng^{H1913}/fg^{wt} PE.

The deletion of these genes might influence the neighboring imprinted genes due to *cis*- and/or *trans*-effects and, as a consequence, expression of downstream genes could be modified (Li et al. 1998). Another possibility would be that improvement in development might lead to altered signaling, which in turn could affect numerous other genes. It is also possible that *H*19 RNA itself might directly regulate expression of a wide range of genes by binding to or interfering with these genes. It is reported that normalization of H19 expression in the ng$^{H193-KO}$/fg^{wt} PE resulted in extended development into E17.5 (Kono et al. 2002) (Fig. 1, Table 1).

*H*19 is expressed abundantly in a broad range of tissues of both endodermal and mesodermal origin in developing mouse embryos (Ohlsson et al. 1994; Svensson et al. 1995). However, the spliced and polyadenylated 2.5-kb *H*19 RNA does not encode a protein, and thus remains untranslated; this RNA species is associated with polysomes in a variety of cells. Several reports have suggested that this RNA plays an active role in tumor suppression (Brannan et al. 1990; Hao et al. 1993). Our analysis demonstrated that expression of not only the imprinted genes located on chromosome 7 but also those located on other chromosomes was normalized in the ng$^{H1913-KO}$/fg^{wt} parthenotes (Kono et al. 2004), suggesting that the reduced level of *H*19 RNA might be involved in the normalization of global gene expression. It should be noted that overexpression of *H*19 does not exert deleterious effects: no obvious phenotype has been detected with an excess dosage of *H*19 in transgenic experiments using a 130-kb YAC clone (Ainscough et al. 1997). An earlier report based on a transgenic study suggested that ectopic expression of

*H*19 can result in prenatal death, but this suggestion was later retracted by the same group. Thus, the roles played by *H*19 remain unclear at present.

In order to gain a better understanding of the synergistic function of multiple genes in development, we focused on another set of coordinately regulated genes, *Dlk1* and *Gtl2*, located on distal chromosome 12, which are paternally imprinted (Schmidt et al. 2000; Kobayashi et al. 2000). These genes were selected for further investigation because mice either with deletion of *Dlk1* or with uniparental disomy for chromosome 12 showed fetal growth retardation and late fetal/neonatal death (Georgiades et al. 2000; Moon et al. 2002). In addition, they shared a number of features with *Igf2* and *H*19 (Takada et al. 2000; Paulsen et al. 2001). A LacZ integration in the *Dlk1-Gtl2* intergenic region leads to proportionate dwarfism when the transgene is inherited from the father (Schuster-Gossler et al. 1996). *Dlk1* is known to encode a transmembrane protein containing epidermal growth factor (EGF) repeats. Moreover, *Dlk1* functions as a growth-promoting factor, since *Dlk1*-null mice displayed fetal growth retardation (Schuster-Gossler et al. 1996). In most ng$^{H1913-KO}$/fg^{wt} parthenotes, *Dlk1* and *Gtl2* were repressed and overexpressed, respectively. Intriguingly, the inappropriate expression pattern was corrected only in the surviving ng$^{H1913-KO}$/fg^{wt} parthenotes. Therefore, it appears that the appropriate expression pattern of these genes was relevant to the normal development of the surviving ng$^{H1913-KO}$/fg^{wt} parthenotes.

3.5 Methylation Analysis of the IG-DMR at Dlk1-Gtl2 Domain

The expressions of *Dlk1, Gtl2, Dio3*, and *Mirg* that are regulated by the methylation status of IG-DMR, located upstream of *Gtl2*, were normally expressed in the survived ng^{H1913}/fg^{wt} PE; this perhaps resulted in normal birth. To gain further insight into the correction of the gene expression, we analyzed the methylation status of IG-DMR of the *Dlk1-Gtl2* domain (Wu et al. 2006). The analysis of the tail of Kaguya showed that IG-DMR of the ng allele was completely methylated and that of the fg allele was also partially methylated. Furthermore, the intestine of another survived ng$^{\text{H1913}}$/fg^{wt} at 19.5 PE also showed complete methylation of the IG-DMR in the ng allele. However, the IG-DMR was completely unmethylated in the tongue and the intestine in the growth-retarded ng$^{\text{H1913}}$/fg^{wt} PE at E18.5. In the controls, the analysis of all tissues exhibited parental allele-specific methylation status with hypermethylated paternal allele and hypomethylated maternal allele. The results thus obtained gave an answer to why *Dlk1* was expressed only in Kaguya, revealing the importance of the normal expression pattern of *imprinted genes located* in the *Dlk1-Gtl2* domain in the normal development and survival of Kaguya. However, the reason for IG-DMR methylation in only a few cases has remained unclear.

3.6 Conclusion

Based on our findings, we propose the paternal dual barrier theory, in which two sets of coordinate imprinted and reciprocally expressed genes, *Igf2-H19* and *Dlk1-Gtl2*, function as a critical barrier to parthenogenetic development in order to render paternal contribution obligatory for the descendants of mammals.

References

Ainscough J, Koide T, Tada M, Barton S, Surani M (1997) Imprinting of Igf2 and H19 from a 130 kb YAC transgene. Development 124:3621–3632

Bao S, Obata Y, Carroll J, Domeki I, Kono T (2000) Epigenetic modifications necessary for normal development are established during oocyte growth in mice. Biol Reprod 62:616–621

Barton SC, Surani MA, Norris ML (1984) Role of paternal and maternal genomes in mouse development. Nature 311:374–376

Brannan CI, Dees EC, Ingram R, Tilghman SM (1990) The product of the H19 gene may function as an RNA. Mol Cell Biol 10:28–36

Feil R, Charlton J, Bird AP, Walter J, Reik W (1994) Methylation analysis on individual chromosomes: improved protocol for bisulphite genomic sequencing. Nucleic Acids Res 22:695–696

Georgiades P, Watkins M, Burton GJ, Ferguson-Smith AC (2001) Roles for genomic imprinting and the zygotic genome in placental development. Proc Natl Acad Sci U S A 98:4522–4527

Hagemann LJ, Peterson AJ, Weilert LL, Lee RS, Tervit HR (1998) In vitro and early in vivo development of sheep gynogenones and putative androgenones. Mol Reprod Dev 50:154–162

Hao Y, Crenshaw T, Moulton T, Newcomb E, Tycko B (1993) Tumour-suppressor activity of H19 RNA. Nature 365:764–767

Hark AT, Schoenherr CJ, Katz DJ, Ingram RS, Levorse JM, Tilghman SM (2000) CTCF mediates methylation-sensitive enhancer-blocking activity at the H19/Igf2 locus. Nature 405:486–489

Hiura H, Obata Y, Komiyama J, Shirai M, Kono T (2006) Oocyte growth-dependent progression of maternal imprinting in mice. Genes Cells, in press

Kobayashi S, Wagatsuma H, Ono R, Ichikawa H, Yamazaki M, Tashiro H, Aisaka K, Miyoshi N, Kohda T, Ogura A, Ohki M, Kaneko-Ishino T, Ishino F (2000) Mouse Peg9/Dlk1 and human PEG9/DLK1 are paternally expressed imprinted genes closely located to the maternally expressed imprinted genes: mouse Meg3/Gtl2 and human MEG3. Genes Cells 5:1029–1037

Kono T, Obata Y, Yoshimzu T, Nakahara T, Carroll J (1996) Epigenetic modifications during oocyte growth correlates with extended parthenogenetic development in the mouse. Nature Genet 13:91–94

Kono T, Sotomaru Y, Katsuzawa Y, Dandolo L (2002) Mouse parthenogenetic embryos with monoallelic H19 expression can develop to day 17.5 of gestation. Dev Biol 243:294–300

Kono T, Obata Y, Wu Q, Niwa K, Ono Y, Yamamoto Y, Park E, Seo J, Ogawa H (2004) Birth of parthenogenetic mice that can develop to adulthood. Nature 428:860–864

Kure-bayashi S, Miyake M, Okada K, Kato S (2000) Successful implantation of in vitro-matured, electro-activated oocytes in the pig. Theriogenology 53:1105–1119

Leighton PA, Ingram RS, Eggenschwiler J, Efstratiadis A, Tilghman SM (1995) Disruption of imprinting caused by deletion of the H19 gene region in mice. Nature 375:34–39

Li YM, Franklin G, Cui HM, Svensson K, He XB, Adam G, Ohlsson R, Pfeifer S (1998) The H19 transcript is associated with polysomes and may regulate IGF2 expression in trans. J Biol Chem 273:28247–28252

Lucifero D, Mann M, Bartolomei M, Trasler J (2004) Gene-specific timing and epigenetic memory in oocyte imprinting. Hum Mol Genet 13:839–849

Mann JR (2001) Imprinting in the germ line. Stem Cells 19:287–294

McGrath J, Solter D (1984) Completion of mouse embryogenesis requires both the maternal and paternal genomes. Cell 37:179–183

Moon YS, Smas CM, Lee K, Villena JA, Kim KH, Yun EJ, Sul HS (2002) Mice lacking paternally expressed Pref-1/Dlk1 display growth retardation and accelerated adiposity. Mol Cell Biol 15:5585–8892

Obata Y, Kono T (2002) Maternal primary imprinting is established at a specific time for each gene throughout oocyte growth. J Biol Chem 277:5285–5289

Obata Y, Kaneko-Ishino T, Koide T, Takai Y, Ueda T, Domeki I, Shiroishi T, Ishino F, Kono T (1998) Disruption of primary imprinting during oocyte growth leads to the modified expression of imprinted genes during embryogenesis. Development 125:1553–1560

Ohlsson R, Hedborg F, Holmgren L, Walsh C, Ekstrom TJ (1994) Overlapping patterns of IGF2 and H19 expression during human development: biallelic IGF2 expression correlates with a lack of H19 expression. Development 120:361–368

Paulsen M, Takada S, Youngson NA, Benchaib M, Charlier C, Segers K, Georges M, Ferguson-Smith AC (2001) Comparative sequence analysis of the imprinted Dlk1-Gtl2 locus in three mammalian species reveals highly conserved genomic elements and refines comparison with the Igf2-H19 region. Genome Res 11:2085–2094

Ripoche MA, Kress C, Poirier F, Dandolo L (1997) Deletion of the H19 transcription unit reveals the existence of a putative imprinting control element. Genes Dev 11:1596–1604

Sasaki H, Ferguson-Smith AC, Shum AS, Barton SC, Surani MA (1995) Temporal and spatial regulation of H19 imprinting in normal and uniparental mouse embryos. Development 121:4195–4202

Schmidt JV, Matteson PG, Jones BK, Guan XJ, Tilghman S (2000) The Dlk1 and Gtl2 genes are linked and reciprocally imprinted. Genes Dev 15:1997–2002

Schuster-Gossler K, Simon-Chazottes D, Guenet JL, Zachgo J, Gossler A (1996) Gtl2lacZ, an insertional mutation on mouse chromosome 12 with parental origin-dependent phenotype. Mamm Genome 7:20–24

Surani MAH, Barton SC (1983) Development of gynogenetic eggs in the mouse: Implications for parthenogenetic embryos. Science 222:1034–1036

Surani MA, Kothary R, Allen ND, Singh PB, Fundele R, Ferguson-Smith AC, Barton SC (1990) Genome imprinting and development in the mouse. Dev Suppl:89–98

Svensson K, Walsh C, Fundele R, Ohlsson R (1995) H19 is imprinted in the choroid plexus and leptomeninges of the mouse foetus. Mech Dev 51:31–37

Takada S, Tevendale M, Baker J, Georgiades P, Campbell E, Freeman T, Johnson MH, Paulsen M, Ferguson-Smith AC (2000) Delta-like and gtl2 are reciprocally expressed, differentially methylated linked imprinted genes on mouse chromosome 12. Curr Biol 10:1135–1138

Tada T, Obata Y, Tada M, Goto Y, Nakatsuji N, Tan S, Kono T, Takagi N (2000) Imprint switching for non-random X-chromosome inactivation during mouse oocyte growth. Development 127:3101–3105

Thorvaldsen JL, Duran KL, Bartolomei MS (1998) Deletion of the H19 differentially methylated domain results in loss of imprinted expression of H19 and Igf2. Genes Dev 12:3693–3702

Wu Q, Kumagai T, Kawahara M, Ogawa H, Hiura H, Obata Y, Takano R, Kono T (2006) Regulated expressions of two sets of paternally imprinted genes are necessary for mouse parthenogenetic development to term. Reproduction, in press

4 Cell-Cell Fusion as a Means to Establish Pluripotency

J.T. Do, H.R. Schöler

4.1   Plasticity of Somatic Stem Cells Versus Cell Fusion 36
4.2   Somatic Cell Reprogramming by Fusion with Embryonic Stem,
        Embryonic Carcinoma, and Embryonic Germ Cells 37
4.3   Epigenetic Modification of Reprogrammed Hybrid Cells 38
4.4   Factors for Fusion-Induced Reprogramming 40
4.5   Perspectives 41
References 42

Abstract. Embryonic stem cells (ESCs), embryonic germ cells (EGCs), and embryonic carcinoma cells (ECCs) are three types of pluripotent cells derived from mammalian embryos. The three cell types are capable not only of self-renewal, but also of having the potential to give rise to cells of all tissue types in the fetal and adult body. In several reports, ESCs, ECCs, and EGCs have been described to reprogram somatic cells in vitro. After reprogramming caused by fusion, somatic cells exhibit various features of pluripotent cells: expression of pluripotency markers (e.g., *Oct4*, *nanog*, and *Rex-1*), absence of tissue-specific gene expression, reactivation of inactive X chromosome of female somatic cells, demethylation, as well as histone modification. An activity in pluripotent stem cells appears to be capable of inducing the global changes inherent in the reprogramming of somatic cells. Investigations involving pluripotent stem cells will yield substantial insight into various fundamental biological processes, such as cellular differentiation and de-differentiation. Most importantly for the public, however, is that such studies might lead into cell-based therapies and as such have the potential to change regenerative medicine.

4.1   Plasticity of Somatic Stem Cells Versus Cell Fusion

Somatic stem cells are capable of self-renewing for long periods of time and of replenishing various cell types lost to tissue injury or other adverse conditions present in their microenvironment (Schofield 1978). Adult stem cells have been found in many types of tissues including the skin (Lavker and Sun 1982; Watt 1998), central nervous system (Gage et al. 1995), muscle (Schultz and McCormick 1994), bone marrow (Weissman 2000), liver (Alison and Sarraf 1998), and mammary glands (Welm et al. 2002). These somatic stem cells preferentially differentiate into cell types consistent with their tissue of origin. For example, neural stem cells differentiate into three major types of neural cells: neurons, oligodendrocytes, and astrocytes. However, many recent studies have suggested that somatic stem cells can contribute to the development of other types of tissues, transgressing the developmental barrier between the cell niche and germ layers. This has been referred to as the plasticity of adult stem cells and has been mainly investigated in hematopoietic and mesenchymal stem cells. Hematopoietic stem cells (HSCs) and mesenchymal stem cells (MSCs) show a high degree of plasticity; they contribute to the development of skin (Krause et al. 2001; Lagasse et al. 2000; Orlic et al. 2001), neurons (Mahmood et al. 2001; Mezey et al. 2000), hepatocytes (Theise et al. 2000), and endothelial cells (Asahara et al. 1999). Even some adult somatic cells exhibit a pluripotent phenotype. Recently, several stem cells, including mesenchymal stem cells (Jiang et al. 2002) and olfactory mucosa stem cells (Murrell et al. 2005), were suggested to be somatic pluripotent cells, because they have a remarkable self-renewal capability as well as the capacity to differentiate spontaneously into all three germ layers: endoderm, mesoderm, and ectoderm. However, until they are tested for their capability to differentiate into germ cells (see Chap. 10, this volume), they are considered multipotent adult stem cells.

Other recent experiments have pointed to the cell fusion between stem cells and somatic cells as the underlying mechanism driving somatic stem cell plasticity (Terada et al. 2002; Ying et al. 2002). Wang et al. (2003) and Vassilopoulos et al. (2003) suggested that hematopoietic stem cells do not have transdifferentiation potential; rather, they simply fuse with hematopoietic cells and function like hematopoietic cells. In addition, Purkinje neurons and cardiac muscle cells were shown contribute to the development of heart, liver, and brain cells by fusion in vivo (Alvarez-Dolado et al. 2003).

4.2 Somatic Cell Reprogramming by Fusion with Embryonic Stem, Embryonic Carcinoma, and Embryonic Germ Cells

It has been reported that somatic cells could be reprogrammed by fusion with EGCs (Tada et al. 1997), ECCs (Flasza et al. 2003; Miller and Ruddle 1976), or ESCs (Do and Schöler 2004, 2005; Hatano et al. 2005; Kimura et al. 2004; Matveeva et al. 1998; Pells et al. 2002; Tada et al. 2001, 2003). When the *Oct4*-GFP transgenic somatic cells were fused with either ESCs (Do and Schöler 2004, 2005; Tada et al. 2001) or EGCs (Tada et al. 1997), GFP-positive cells were observed, suggesting that the somatic cell genome itself could be reprogrammed. Because the *Oct4* gene is active in ESCs, but not in somatic cells, it follows that *Oct4*-GFP can only be expressed in reprogrammed hybrid cells. To enhance our understanding of somatic cell-genome reactivation, inter-species hybridization experiments between the pluripotent mouse P19 ECC and the human T lymphoma cell line CEM-GFP were conducted (Flasza et al. 2003). The subsequent hybrid cells expressed both human and mouse pluripotent markers: *hOct4*, *hSox-2*, *mOct4*, and *mSox-2*. On the other hand, somatic cell-specific genes were inactivated after fusion with ESCs. The neural stem cell marker *Nestin* and the ectoderm marker *glutamate receptor 6* (*Glur6*), which are expressed in neuro-sphere cells, are not expressed after fusion with ESCs (Do and Schöler 2005). Therefore, both the reactivation of pluripotent-specific marker genes and the inactivation of tissue-specific genes are common features in fusion-induced reprogramming.

To determine the developmental potential of the hybrid cells, they were injected into blastocysts and allowed to develop. Following the injection of EG-thymocyte hybrid cells into blastocysts, these cells are found to contribute to the development of embryonic ectoderm, visceral endoderm, yolk-sac mesoderm, and embryonic mesoderm in 7.5–10.5 days postcoitus (dpc) embryos (Tada et al. 1997, 2001). Even the ES-splenocyte hybrid cells exhibit a full-term contribution to chimeras (Matveeva et al. 1998).

4.3 Epigenetic Modification of Reprogrammed Hybrid Cells

Cloning experiments have shown that the epigenetic marks of donor cells are capable of being reestablished during reprogramming (Eggan et al. 2000). The three pluripotent cells (ESCs, EGCs, and ECCs) can form hybrid cells by fusion with somatic cells and can induce epigenetic changes – such as X chromosome inactivation, DNA methylation, and histone modification – by which gene expression patterns are altered from the differentiated to the undifferentiated state (Fig. 1). For example, the genes on one of the two X chromosomes in female somatic cells become transcriptionally inactive as a result of X chromosome inactivation (Lyon 1999). Both X chromosomes are active in the preimplantation stage of the embryo and in undifferentiated female ESCs (Lyon 1999), ECCs (McBurney and Strutt 1980), and EGCs (Stewart et al. 1994); however, only one of the X chromosomes is inactive in somatic cells.

Reactivation of the inactive X chromosome of female somatic cells can be induced by cell fusion with ECCs (Takagi et al. 1983). This is evidenced, for example, by male ES-XX thymocyte hybrid cells, which carry three synchronously replicating X chromosomes, indicating that the inactive X chromosome of thymocytes must have been reactivated (Tada et al. 2001). Also, the inactive X chromosome of lymphocytes was reactivated by adopting the unstable *Xist* expression pattern of EGCs (Nesterova et al. 2002). In addition, the expression of *Tsix*, an antisense RNA that binds to and regulates *Xist* (Lee and Lu 1999), also reversed during the fusion-induced reprogramming (Kimura et al. 2002; Nesterova et al. 2002).

The change in the methylation patterns of imprinted versus nonimprinted genes has also been examined during EGC fusion-induced reprogramming. Tada et al. (1997) reported that EGCs induce demethylation

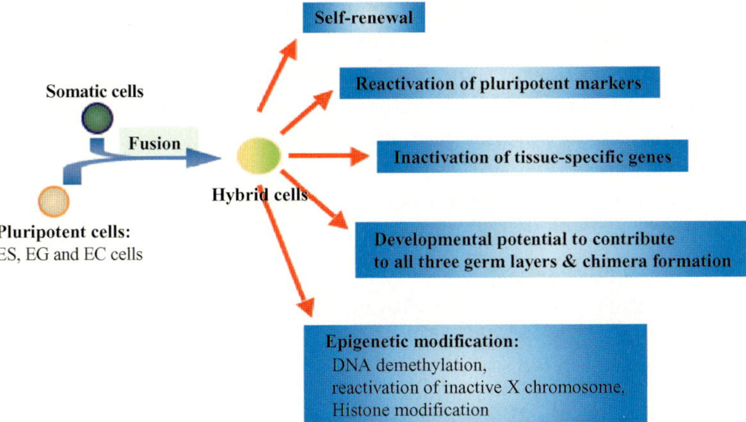

Fig. 1. Established pluripotency of somatic genome through cell fusion. The genome of somatic cells can be successfully reprogrammed by forming cell hybrids with any of the three pluripotent cells: embryonic stem (*ES*) cells, embryonic germ (*EG*), and embryonic carcinoma (*EC*) cells. By this process, they acquire pluripotential characteristics, such as self-renewal, reactivation of pluripotent-specific marker genes, inactivation of tissue-specific genes, developmental potential to contribute to all three germ layers, and epigenetic modification to undifferentiated

of somatic cells when fused with somatic cells. Moreover, demethylation of maternally imprinted genes (*Igf2*, *Peg3*, and *Peg1/Mest*), and nonimprinted genes (*Aprt*, *Pgk-2*, and β *globin*) could induce appropriate expression of the methylated silent allele (Tada et al. 1997). However, unlike EGCs, ESCs were not capable of inducing a change in the methylation status of imprinted genes when they were fused with thymocytes (Tada et al. 2001). The authors suggested that EGCs assume an additional dominant potential to induce more extensive demethylation of somatic cells.

Recently, Kimura et al. (2004) showed that ESCs were capable of inducing histone modifications on the genome of reprogrammed somatic cells through hybridization (Kimura et al. 2004). The histone modification, such as acetylation and methylation of lysine positions of histone H3 (H3) and H4, is thought to play a key role in regulating gene expression (Lachner et al. 2003). Histone acetylation correlates with the active gene state, while deacetylation correlates with the silent gene state. The higher acetylation of H3 and H4 and di- and trimethylation of H3-K4 (lysine 4 of histone H3) were examined in fusion hybrid cells (Kimura et al. 2004). Collectively, these data suggest that the three pluripotent cell types have an intrinsic capability of inducing the global epigenetic changes during the hybrid-induced reprogramming of somatic cells.

4.4   Factors for Fusion-Induced Reprogramming

Cell fusion technology is a powerful tool that has the potential to uncover the specific factors contributing to the reprogramming process. In our recent report, we sought to determine the source in ESCs that is responsible for the reprogramming of somatic cells. To this end, we fused mouse neural cells with either ESC cytoplasts or ESC karyoplasts. These experiments showed that the karyoplasts of ESCs could induce the expression of the pluripotent markers *Oct4*, *Rex 1*, and *nanog* of somatic genome, but the cytoplasts failed to reprogram Oct4-GFP expression of neural cells after fusion (Do and Schöler 2004). Taken together, these findings show that ESC nuclei either contain all the factors necessary to reprogram somatic cells, or sufficient quantities thereof; the cytoplasm, however, lacks factors crucial for reactivation, or contains insufficient levels of existing factors. Ringertz et al. also suggested that nuclear factors are crucial for reactivation of the cell genome in fusion hybrids (Ringertz et al. 1971). In the cloning procedure through nuclear transfer using oocytes, it has been suggested that the nuclear factors are essential for nuclear remodeling (Gao et al. 2002) and successful full-term development (Wakayama et al. 2000). In addition, our investigations led us to suggest that Oct4 reprogramming ability is independent of DNA replication and cell division.

It is known that the less specialized cells are, the more readily they can be reprogrammed during cloning procedures. If the same holds true in fusion experiments, the successful reprogramming of somatic stem cells may be induced with higher efficiency than that for differentiated somatic cells. However, we could not find any difference in the reprogramming efficiency of the somatic stem cell-containing cell population (neurosphere cells) and the differentiated cell population (cumulus cells). Instead, with respect to reprogramming lag time, somatic gene expression changes, reprogramming rate, and in vitro differentiation potential, we found that the reprogramming efficiency of cumulus cells was comparable to that of neurosphere cells (Do and Schöler 2005).

4.5 Perspectives

Therapeutic cloning, which can provide immune-matched ESCs, has been suggested as a means for tissue replacement therapy. However, this method has raised a number of controversies, as it requires that women donate oocytes and that embryos are generated for the sake of deriving ESC lines. Recently, several alternative methods have been suggested, including in vitro oocyte derivation from ESCs (Hübner et al. 2003), induction of dedifferentiation of somatic cells by fusion with ESCs (Do and Schöler 2004), and chemically induced reprogramming (Ding and Schultz 2004). A promising alternative may be fusion-induced reprogramming. As we discussed above, fusion of somatic cells with ESCs induces successful reprogramming of the somatic genome. If the ESC genome is removed from the hybrids to eliminate a potential immune response, immune-matched reprogrammed cells from a patient's own cells could be obtained. It is possible that ESC cytoplasts obtained after removal of chromosomes from the metaphase stage still contain the nuclear factors that reprogram the somatic nucleus. Finally, it remains incumbent on researchers in tissue replacement therapy to determine the nature of reprogramming factors, with which somatic cells can be dedifferentiated and redifferentiated into preselected tissue for biomedical applications. In this respect, it is encouraging that results obtained with mouse also apply to human ESCs (Cowan et al. 2005).

Acknowledgements. This work was supported by the BMBF grant "Reprogramming of Somatic Cells for the Therapy of Heart Disorders" (reference number: 01GN0539).

References

Alison M, Sarraf C (1998) Hepatic stem cells. J Hepatol 29:676–682

Alvarez-Dolado M, Pardal R, Garcia-Verdugo JM, Fike JR, Lee HO, Pfeffer K, Lois C, Morrison SJ, Alvarez-Buylla A (2003) Fusion of bone-marrow-derived cells with Purkinje neurons, cardiomyocytes and hepatocytes. Nature 425:968–973

Asahara T, Masuda H, Takahashi T, Kalka C, Pastore C, Silver M, Kearne M, Magner M, Isner JM (1999) Bone marrow origin of endothelial progenitor cells responsible for postnatal vasculogenesis in physiological and pathological neovascularization. Circ Res 85:221–228

Cowan CA, Atienza J, Melton DA, Eggan K (2005) Nuclear reprogramming of somatic cells after fusion with human embryonic stem cells. Science 309:1369–1373

Ding S, Schultz PG (2004) A role for chemistry in stem cell biology. Nat Biotechnol 22:833–840

Do JT, Schöler HR (2004) Nuclei of embryonic stem cells reprogram somatic cells. Stem Cells 22:941–949

Do JT, Schöler HR (2005) Comparison of neurosphere cells with cumulus cells after fusion with embryonic stem cells: reprogramming potential. Reprod Fertil Dev 17:143–149

Eggan K, Akutsu H, Hochedlinger K, Rideout W 3rd, Yanagimachi R, Jaenisch R (2000) X-Chromosome inactivation in cloned mouse embryos. Science 290:1578–1581

Flasza M, Shering AF, Smith K, Andrews PW, Talley P, Johnson PA (2003) Reprogramming in inter-species embryonal carcinoma-somatic cell hybrids induces expression of pluripotency and differentiation markers. Cloning Stem Cells 5:339–354

Gage FH, Coates PW, Palmer TD, Kuhn HG, Fisher LJ, Suhonen JO, Peterson DA, Suhr ST, Ray J (1995) Survival and differentiation of adult neuronal progenitor cells transplanted to the adult brain. Proc Natl Acad Sci U S A 92:11879–11883

Gao S, Gasparrini B, McGarry M, Ferrier T, Fletcher J, Harkness L, De Sousa P, Wilmut I (2002) Germinal vesicle material is essential for nucleus remodeling after nuclear transfer. Biol Reprod 67:928–934

Hatano SY, Tada M, Kimura H, Yamaguchi S, Kono T, Nakano T, Suemori H, Nakatsuji N, Tada T (2005) Pluripotential competence of cells associated with Nanog activity. Mech Dev 122:67–79

Hübner K, Fuhrmann G, Christenson LK, Kehler J, Reinbold R, De La Fuente R, Wood J, Strauss JF 3rd, Boiani M, Schöler HR (2003) Derivation of oocytes from mouse embryonic stem cells. Science 300:1251–1256

Jiang Y, Jahagirdar BN, Reinhardt RL, Schwartz RE, Keenek CD, Ortiz-Gonzalezk XR, Reyes M, Lenvik T, Lund T, Blackstad M, Du J, Aldrich S, Lisberg A, Lowk WC, Largaespada DA, Verfaillie CM (2002) Pluripotency of mesenchymal stem cells derived from adult marrow, Nature 418:41–19

Kimura H, Tada M, Hatano S, Yamazaki M, Nakatsuji N, Tada T (2002) Chromatin reprogramming of male somatic cell-derived XIST and TSIX in ES hybrid cells. Cytogenet Genome Res 9:106–114

Kimura H, Tada M, Nakatsuji N, Tada T (2004) Histone code modifications on pluripotential nuclei of reprogrammed somatic cells. Mol Cell Biol 24:5710–5720

Krause DS, Theise ND, Collector MI, Henegariu O, Hwang S, Gardner R, Neutzel S, Sharkis SJ (2001) Multi-organ, multi-lineage engraftment by a single bone marrow-derived stem cell. Cell 105:369–377

Lachner M, O'Sullivan RJ, Jenuwein T (2003) An epigenetic road map for histone lysine methylation. J Cell Sci 116:2117–2124

Lagasse E, Connors H, Al-Dhalimy M, Reitsma M, Dohse M, Osborne L, Wang X, Finegold M, Weissman IL, Grompe M (2000) Purified hematopoietic stem cells can differentiate into hepatocytes in vivo. Nat Med 6:1229–1234

Lavker RM, Sun TT (1982) Heterogeneity in epidermal basal keratinocytes: morphological and functional correlations. Science 215:1239–1241

Lee JT, Lu N (1999) Targeted mutagenesis of Tsix leads to nonrandom X inactivation. Cell 99:47–57

Lyon MF (1999) X-chromosome inactivation. Curr Biol 9:R235–R237

Mahmood A, Lu D, Wang L, Li Y, Lu M, Chopp M (2001) Treatment of traumatic brain injury in female rats with intravenous administration of bone marrow stromal cells. Neurosurgery 49:1196–1203; discussion 1203–1204

Matveeva NM, Shilov AG, Kaftanovskaya EM, Maximovsky LP, Zhelezova AI, Golubitsa AN, Bayborodin SI, Fokina MM, Serov OL (1998) In vitro and in vivo study of pluripotency in intraspecific hybrid cells obtained by fusion of murine embryonic stem cells with splenocytes. Mol Reprod Dev 50:128–138

McBurney MW, Strutt BJ (1980) Genetic activity of X chromosomes in pluripotent female teratocarcinoma cells and their differentiated progeny. Cell 21:357–364

Mezey E, Chandross KJ, Harta G, Maki RA, McKercher SR (2000) Turning blood into brain: cells bearing neuronal antigens generated in vivo from bone marrow. Science 290:1779–1782

Miller RA, Ruddle FH (1976) Pluripotent teratocarcinoma-thymus somatic cell hybrids. Cell 9:45–55

Murrell W, Feron F, Wetzig A, Cameron N, Splatt K, Bellette B, Bianco J, Perry C, Lee G, Mackay-Sim A (2005) Multipotent stem cells from adult olfactory mucosa. Dev Dyn 233:496–515

Nesterova TB, Mermoud JE, Hilton K, Pehrson J, Surani MA, McLaren A, Brockdorff N (2002) Xist expression and macroH2A1.2 localisation in mouse primordial and pluripotent embryonic germ cells. Differentiation 69:216–225

Orlic D, Kajstura J, Chimenti S, Jakoniuk I, Anderson SM, Li B, Pickel J, McKay R, Nadal-Ginard B, Bodine DM, Leri A, Anversa P (2001) Bone marrow cells regenerate infarcted myocardium. Nature 410:701–705

Pells S, Di Domenico AI, Gallagher EJ, McWhir J (2002) Multipotentiality of neuronal cells after spontaneous fusion with embryonic stem cells and nuclear reprogramming in vitro. Cloning Stem Cells 4:331–338

Ringertz NR, Carlsson SA, Ege T, Bolund L (1971) Detection of human and chick nuclear antigens in nuclei of chick erythrocytes during reactivation in heterokaryons with HeLa cells. Proc Natl Acad Sci U S A 68:3228–3232

Schofield R (1978) The relationship between the spleen colony-forming cell and the haemopoietic stem cell. Blood Cells 4:7–25

Schultz E, McCormick KM (1994) Skeletal muscle satellite cells. Rev Physiol Biochem Pharmacol 123:213–257

Stewart CL, Gadi I, Bhatt H (1994) Stem cells from primordial germ cells can reenter the germ line. Dev Biol 161:626–628

Tada M, Tada T, Lefebvre L, Barton SC, Surani MA (1997) Embryonic germ cells induce epigenetic reprogramming of somatic nucleus in hybrid cells. EMBO J 16:6510–6520

Tada M, Takahama Y, Abe K, Nakatsuji N, Tada T (2001) Nuclear reprogramming of somatic cells by in vitro hybridization with ESCs. Curr Biol 11:1553–1558

Tada M, Morizane A, Kimura H, Kawasaki H, Ainscough JF, Sasai Y, Nakatsuji N, Tada T (2003) Pluripotency of reprogrammed somatic genomes in embryonic stem hybrid cells. Dev Dyn 227:504–510

Takagi N, Yoshida MA, Sugawara O, Sasaki M (1983) Reversal of X-inactivation in female mouse somatic cells hybridized with murine teratocarcinoma stem cells in vitro. Cell 34:1053–1062

Terada N, Hamazaki T, Oka M, Hoki M, Mastalerz DM, Nakano Y, Meyer EM, Morel L, Petersen BE, Scott EW (2002) Bone marrow cells adopt the phenotype of other cells by spontaneous cell fusion. Nature 416:542–545

Theise ND, Badve S, Saxena R, Henegariu O, Sell S, Crawford JM, Krause DS (2000) Derivation of hepatocytes from bone marrow cells in mice after radiation-induced myeloablation. Hepatology 31:235–240

Vassilopoulos G, Wang PR, Russell DW (2003) Transplanted bone marrow regenerates liver by cell fusion. Nature 422:901–904

Wakayama T, Tateno H, Mombaerts P, Yanagimachi R (2000) Nuclear transfer into mouse zygotes. Nat Genet 24:108–109

Wang X, Willenbring H, Akkari Y, Torimaru Y, Foster M, Al-Dhalimy M, Lagasse E, Finegold M, Olson S, Grompe M (2003) Cell fusion is the principal source of bone-marrow-derived hepatocytes. Nature 422:897–901

Watt FM (1998) Epidermal stem cells: markers, patterning and the control of stem cell fate. Philos Trans R Soc Lond B Biol Sci 353:831–837

Weissman IL (2000) Translating stem and progenitor cell biology to the clinic: barriers and opportunities. Science 287(5457):1442–1446

Welm BE, Tepera SB, Venezia T, Graubert TA, Rosen JM, Goodell MA (2002) Sca-1(pos) cells in the mouse mammary gland represent an enriched progenitor cell population. Dev Biol 245:42–56

Ying QL, Nichols J, Evans EP, Smith AG (2002) Changing potency by spontaneous fusion. Nature 416:545–548

5 Toward Reprogramming Cells to Pluripotency

P. Collas, C.K. Taranger

5.1   Introduction   48

5.2   Epigenetic Reprogramming and Induction of Pluripotency
       by Cell Fusion   49

5.2.1 Epigenetic Reprogramming in Hybrids of Somatic Cells
       with Embryonic Germ Cells   49

5.2.2 Nuclear Reprogramming in Somatic Cells Fused
       to Embryonic Stem or Embryonic Carcinoma Cells   50

5.3   Induction of Differentiation, Transdifferentiation,
       and Dedifferentiation Using Cell Extracts   50

5.3.1 Stem Cell Differentiation and Somatic Cell Transdifferentiation . .   51

5.3.2 Induction of Cellular Dedifferentiation with Extracts
       of Blastema and Eggs   53

5.3.3 Nuclear Reprogramming with Extract
       of Undifferentiated Carcinoma Cells   54

5.3.4 How Plastic Are Extract-Treated Cells?   60

5.4   Chromatin Remodeling Associated with Nuclear Reprogramming .   61

5.5   Perspectives   62

5.5.1 Innovative Aspects of Extract-Mediated Cell Reprogramming . . .   62

5.5.2 Cell Extracts as a Tool for Investigating
       Nuclear Reprogramming Mechanisms   63

References   64

Abstract. The possibility of turning one somatic cell type into another may in the long run have beneficial applications in regenerative medicine. Somatic cell nuclear transfer (therapeutic cloning) may offer this possibility; however, ethical guidelines prevent application of this technology in many in countries. As a result, alternative approaches are being developed for altering cell fate. This communication discusses recent non-nuclear transfer-based in vitro approaches for reprogramming cells and enhancing their potential for differentiation toward various lineages.

5.1 Introduction

Although under normal conditions the fate of a differentiated cell is generally stable, there is natural and experimental evidence to indicate that dedifferentiation events can take place. A classical illustration is the replacement of lost anatomical parts in teleost fish and urodele amphibians. There, a limb, for example, can be regrown by a process of migration, dedifferentiation, proliferation, and redifferentiation of epithelial cells in the wounded area. Experimentally, a paradigm of dedifferentiation is the functional reprogramming of a differentiated cell nucleus by transplantation into an unfertilized oocyte or egg. Somatic cell nuclear transfer has been shown to result in the derivation of pluripotent embryonic stem cells (ESCs) from cloned blastocyst-stage embryos (Cibelli et al. 1998; Munsie et al. 2000; Wakayama et al. 2001) and in the live birth of cloned offspring in several species (Wilmut et al. 2002; Gurdon and Byrne 2003). The mechanisms of nuclear reprogramming, at least as it occurs by nuclear transfer, have started to be elucidated and point to a process requiring DNA demethylation (Simonsson and Gurdon 2004).

While somatic cell nuclear transfer was being developed, cell fusion approaches led to the demonstration that a differentiated somatic cell nucleus could be epigenetically reprogrammed to express embryonic genes (Tada et al. 1997, 2001). Additionally, hybridization of bone marrow-derived stem cells or neuronal stem cells with ESCs changes the potency of the somatic cells (Terada et al. 2002; Ying et al. 2002). Furthermore, fusions with embryonic germ cells (EGCs), ESCs, or embryonal carcinoma cells (ECCs) have identified differential reprogramming properties between these undifferentiated cell types.

Reprogramming of nuclei by nuclear transplantation and cell hybridization has provided a rationale for the emergence of cell-free (in vitro) strategies for triggering a new differentiation program in stem cells or in already differentiated cells. These approaches rely on the use of an extract from a chosen "target" cell type (the cell type one wishes to turn a cell into). The extract is expected to contain regulatory components necessary for driving the fate of one given cell type into that of the target cell type (Collas and Håkelien 2003). Cell extract-based systems for reprogramming cell fate have been developed with the aim of eliciting somatic cell transdifferentiation (Håkelien et al. 2002), dedifferentiation (McGann et al. 2001; Hansis et al. 2004), or stem cell differentiation (Gaustad et al. 2004; Qin et al. 2005).

5.2 Epigenetic Reprogramming and Induction of Pluripotency by Cell Fusion

5.2.1 Epigenetic Reprogramming in Hybrids of Somatic Cells with Embryonic Germ Cells

Elegant cell fusion studies involving mouse thymocytes with EGCs or ESCs have shown that various degrees of epigenetic reprogramming could be triggered in the thymocyte nuclei (Tada et al. 1997, 2001). EGCs are derived from primordial gem cells (PGCs) and harbor a similar epigenome to PGCs. Because PGCs undergo a complete epigenetic reprogramming during development (through genome-wide DNA demethylation), the reprogramming ability of EGCs has been investigated (Tada et al. 1997). Interestingly, EG-thymocyte hybrids were characterized by demethylation of several imprinted and nonimprinted genes. These alterations were heritable and affected transcription, as shown by the reactivation of the silent maternal allele of the *Peg1/Mest* imprinted gene in the thymocyte nucleus (Tada et al. 1997). Epigenetic changes in the thymocyte nucleus were consistent with induction of pluripotency markers in the hybrid cells, such as their ability to differentiate into all three germ layers in chimeric embryos (Tada et al. 1997).

5.2.2 Nuclear Reprogramming in Somatic Cells Fused to Embryonic Stem or Embryonic Carcinoma Cells

ES cells can also promote some molecular markers of reprogramming such as thymocyte-derived X chromosome reactivation (Tada et al. 2001). Like EGCs, ESCs can induce pluripotency in somatic cells including thymocytes (Tada et al. 2001), neuronal progenitor cells (Ying et al. 2002; Pells et al. 2002), or bone marrow-derived cells (Terada et al. 2002), and are compatible with induction of all germ layers in chimeric mice (Tada et al. 2001; Terada et al. 2002) or in teratomas (Ying et al. 2002). Similarly, fusion of ECCs with T lymphoma cells promotes the formation of colonies expressing pluripotent cell transcripts from the lymphoma cell genome (Flasza et al. 2003). Collectively, this work argues that factors in undifferentiated cells can elicit some epigenetic reprogramming, with associated changes in gene expression, in a more differentiated or committed cell type.

Interestingly however, in contrast to EGC–thymocyte hybrids, ESC–thymocyte hybrids do not support DNA demethylation of imprinted genes, although the hybrids are capable of activating (and therefore presumably demethylating; Simonsson and Gurdon 2004) an *Oct4-GFP* transgene (Tada et al. 2001). Therefore, although ESCs can reset some aspects of epigenetic reprogramming in differentiated cell nuclei, it seems at present that the ability to fully reprogram the epigenome, and potentially gene expression, in differentiated nuclei is restricted to EGCs.

5.3 Induction of Differentiation, Transdifferentiation, and Dedifferentiation Using Cell Extracts

The cell hybridization studies mentioned above have provided a rationale for the development of nuclear and cytoplasmic extracts aimed at reprogramming cells. The principle is simple: when introduced into a recipient cell, an extract from a chosen target cell type would be expected to provide regulatory components necessary to impose a nuclear program characteristic of the target cell type. This approach has been developed by us and others with varying success depending on the recipient and target cell types investigated.

5.3.1 Stem Cell Differentiation and Somatic Cell Transdifferentiation

With the idea of transdifferentiating somatic cells, our laboratory has developed nuclear and cytoplasmic extracts from several model target cell types (Håkelien et al. 2002) (Fig. 1). The procedure entails reversible permeabilization of recipient cells (e.g., kidney epithelial 293T cells) with the pore-forming toxin streptolysin O (SLO), exposure of the permeabilized cells to the extract, $CaCl_2$-mediated resealing of the cells, followed by culture.

Using this approach, we have shown that 293T cells treated with an extract of Jurkat T cells take on T cell properties. These include histone H4 hyperacetylation, an indicator of gene activity, at a locus specifically active in T cells (the *interleukin 2* (*IL2*) locus; Landsverk et al. 2002), expression of T cell-specific genes including *IL2* and surface receptors, and induction of T cell-specific intracellular signaling pathways (assembly of the IL-2 receptor and IL-2 hormone secretion) (Håkelien et al. 2002; Håkelien et al. 2005). Some of the new phenotypes appear stable for several months in culture; however, there are no data to date to indicate how stable the changes would be in vivo. To corroborate these findings in another system, a related approach has illustrated the induction of the pancreas-specific genes *Pdx1* and *Insulin* in rat primary fibroblasts treated with an extract of rat insulinoma cells (Håkelien et al. 2004).

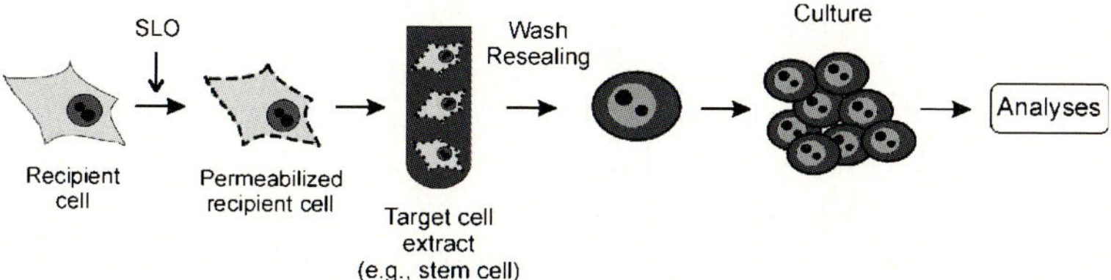

Fig. 1. Principle of reprogramming permeabilized cells using cell extracts. Recipient cells are reversibly permeabilized with streptolysin O (*SLO*), exposed to a nuclear and cytoplasmic extract of a chose target cell type, resealed and cultured for analysis

Cell extracts can also elicit signs of differentiation in somatic or embryonic stem cells. Lysates of fetal rat cardiomyocytes were shown to promote expression of cardiomyocyte proteins and functions in human adipose stem cells (Gaustad et al. 2004). These markers included sarcomeric α-actinin, troponin I, desmin, and, in less than 2% of the cells, beating in culture. More recently, a pneumocyte extract was shown to induce differentiation of mouse ESCs to express a pneumocyte-specific transgene and take on a pneumocyte phenotype (Qin et al. 2005). Whether cell extracts provide more efficient ways of inducing stem cell differentiation remains to be shown; however, the findings of Qin and colleagues seem to suggest so, at least in their system.

Because these studies rely on in vitro characterization of extract-treated cells, the extent to which the novel phenotype – or the new program – is stable remains uncertain. Some studies indicate that the changes elicited are transient, with the duration of expression of a target cell-specific gene or phenotype varying with the type of cell to be reprogrammed and/or the source of extract – although these parameters are by no means exhaustive. For example, expression of *Pdx1* or *Insulin* genes in fibroblasts treated with insulinoma extract lasts for 1–3 weeks, although it may be first detected as late as several weeks after extract exposure (Håkelien et al. 2004). Furthermore, a recent microarray analysis of gene expression in 293T cells exposed to a Jurkat T cell extract indicates that among Jurkat cell-specific genes induced or upregulated within 1 week of extract treatment, only a fraction ($<20\%$) remains expressed after 3 months (Håkelien et al. 2005). Similarly, some genes downregulated by the extract gradually become reactivated over time. Whether these variations reflect an incomplete reprogramming of gene expression, the deregulation of a program properly initiated (owing to, for example, suboptimal culture conditions), or conversely, the dynamics of transcriptional changes necessary to establish a stable phenotype remains at present unresolved.

Several approaches may enhance the stability of extract-induced gene expression, such as repeated extract exposure and induction of a dedifferentiation step prior to triggering a new target cell-specific program of gene expression. The latter strategy is currently being explored (see below).

5.3.2 Induction of Cellular Dedifferentiation with Extracts of Blastema and Eggs

The successful production of ESCs and offspring by nuclear transplantation (Cibelli et al. 1998; Wilmut et al. 2002; Gurdon and Byrne 2003), together with the activation and DNA demethylation of embryonic genes in somatic-EGC or ESC hybrids (Tada et al. 1997, 2001), indicate that undifferentiated cells contain regulatory factors necessary for eliciting pluripotency in a differentiated cell. Induction of dedifferentiation on a large scale would benefit from the development of cellular extracts capable of treating millions of nuclei or cells simultaneously. Several recent lines of evidence suggest that cell extracts may be useful for reprogramming cells to pluripotency.

A first illustration is based on the induction of dedifferentiation with extracts of regenerating newt limbs (McGann et al. 2001). When continuously exposed to cultured differentiated (postmitotic) mouse C2C12 myotubes for several days, regenerating limb extracts promote cell cycle reentry in 18% of the myotubes, as shown by induction of DNA synthesis (McGann et al. 2001). This is accompanied by a downregulation of muscle-specific markers in 15%–30% of the myotubes. Interestingly, mitosis is also detected in roughly 10% of the myotubes and approximately half of these continue proliferating as mononucleated cells in culture (McGann et al. 2001). Interestingly, these findings imply that the dedifferentiated phenotype is maintained even after removal of the extract, suggesting that some (possibly long-term) functional reprogramming events have taken place. It would be informative to characterize the response of other terminally differentiated cell types to extract from regenerating limbs or blastema.

As anticipated from recent nuclear transplantation work (Byrne et al. 2003; Simonsson and Gurdon 2004), extracts of *Xenopus* eggs can also induce expression of pluripotency markers in 293T cells and primary leukocytes (Hansis et al. 2004). The cells form expanding clusters resembling ESC colonies, and upregulate expression of *OCT4* and *germ cell alkaline phosphatase* (*GCAP*) while downregulating differentiation markers (Hansis et al. 2004). However, reprogrammed leukocytes have been shown to have a limited life span and do not express surface markers characteristic of stem cells, indicating that, as observed with extract from other cell types (see above), reprogramming under these conditions is only partial. In light of these findings, lysates derived from mammalian oocytes could probably reprogram somatic nuclei. This hypothesis remains to be tested, but this remains a technical challenge due to the relatively limited amount of material obtainable from the minuscule mammalian oocytes compared to those of amphibians.

5.3.3 Nuclear Reprogramming with Extract of Undifferentiated Carcinoma Cells

Our laboratory has recently extended the above studies to show that nuclear and cytoplasmic extracts of undifferentiated human teratocarcinoma cells can reprogram 293T cells to take on properties of undifferentiated cells with a potential for pluripotency (Taranger et al. manuscript submitted). Teratocarcinomas are a particular type of germ cell tumor that contain undifferentiated stem cells and differentiated derivatives that can include endoderm, mesoderm, and ectoderm germ layers (Chambers and Smith 2004). Undifferentiated ECCs form malignant teratocarcinomas when transplanted into ectopic sites. Interestingly however, some ECC lines can also contribute to tissues of a developing fetus when introduced into a blastocyst (Blelloch et al. 2004).

Phenotypic Characteristics of Reprogrammed Cells

A first result of treatment with EC (NCCIT cell line) extract is a change in cell morphology. Colonies of 293T cells develop, contained within defined edges, and are maintained for many passages in culture (Fig. 2). Notably, cells exposed to their own extract or to extract of Jurkat T cells do not form colonies or form clearly morphologically distinct aggregates (Håkelien et al. 2005) (Fig. 2).

A second line of evidence of reprogramming by EC extract is the induction of Oct4 gene and protein expression. We found that 60%–70% of the cells treated with EC extract display intranuclear Oct4 protein labeling 1 week after EC extract treatment (Fig. 3A). Concomitantly, nuclear lamin A/C, a differentiated cell marker (Hutchison and Worman 2004) was strongly downregulated – a clear sign of dedifferentiation (Fig. 3B). Induction of *OCT4* (*POU5F1*) transcription and loss

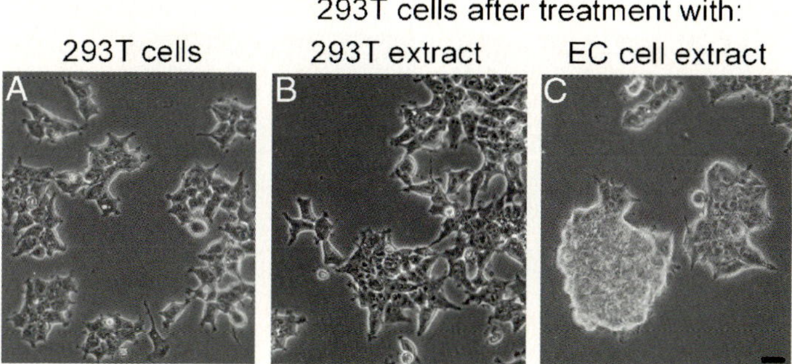

Fig. 2. A–C Changes in morphology of 293T cells treated with ECC extract. **A** Untreated, input 293T cells. **B** 293T cells treated with extract of 293T or **C** ECCs. Bar, 20 μm

of *LMNA* (lamin A) gene expression over time were also quantifiably evident (Fig. 3C). Furthermore, *OCT4* expression was accompanied by expression of additional markers of pluripotency including the Oct4-responsive gene *SOX2* (Fig. 3C) (see also below).

Transcriptional Changes

Gene expression profiling of EC extract-treated cells showed that roughly 1,800 and approximately 1,700 genes were up- and downregulated, respectively, relative to 293T cells. Of these, roughly 70% and approximately 34%, respectively, were shared with ECCs (EC genes). Interestingly, exposure of 293T cells to their own extract significantly altered expression of only a handful of EC genes ($\sim$ 5% of total altered genes), as did an extract of Jurkat T cells (Håkelien et al. 2005). This observation argues for some target cell type specificity in the transcriptional changes elicited. Further, essentially all EC genes affected by 293T or Jurkat extract were the same and statistical analysis indicated that these genes were altered by chance rather than by extract treatment.

When we examined the consistency of EC gene expression changes in EC extract-treated cells, we found that approximately 700 genes immediately upregulated or induced remained expressed for over 2 months, while over 160 genes remained consistently downregulated. Most an-

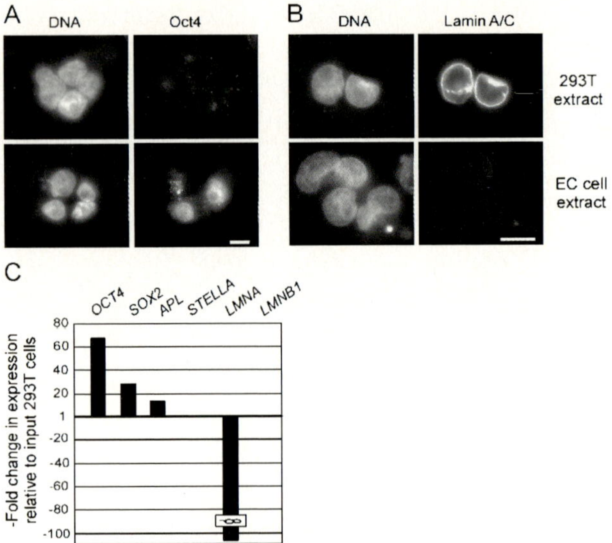

Fig. 3. A–C Induction of Oct4 expression and downregulation of lamin A/C expression in 293T cells reprogrammed in ECC extract. **A** Intranuclear immunolabeling of Oct4 protein in 293T cells treated with control 293T extract or EC extract. **B** Disappearance of perinuclear lamin A/C labeling in EC extract, but not 293T extract-treated cells. **C** Quantitative RT-PCR analysis of expression of indicated genes in cells exposed to ECC extract relative to expression levels in cells exposed to 293T extract. Expression levels were adjusted to those of *GAPDH*. Bars, 10 μm

notated upregulated genes encoded elements involved in transcription, cytoskeletal organization, metabolism, signaling, and chromatin remodeling, whereas downregulated genes were more evenly distributed across classes.

These observations suggest some target cell-type specificity in the changes of gene expression and that some of the alterations are stable (at least for a few months). Yet it is also clear that probably not all changes are heritable. Genes with unstable expression pattern may include "passive bystanders," which generate a transcriptional noise (Paulsson 2004), resulting from more specific alterations in the transcriptional network. Perturbation in the network, however, would be expected to induce changes trickling down the network until a transcriptional equilibrium is reached (Miklos and Maleszka 2004). Accordingly, expression of these genes would be expected to return to a background level; however, we have not noticed such stabilization, suggesting that, in addition to target cell type-specific changes, long-lasting (and as yet uncharacterized) perturbations in the transcriptome are also triggered. To date, these alterations are not fully controlled. Fluctuations in the gene expression profile may result from incomplete reprogramming and from heterogeneity in the transcriptional response to extracts. Nonetheless, the dynamics of gene expression may also illustrate a temporal compartmentalization, in terms of timing, duration, and periodicity of gene activity required to establish a heritable transcriptional network (Klevecz et al. 2004).

Evidence of Induction of Potential for Pluripotency and Multilineage Priming

Downregulation of Differentiated Cell-Specific Genes   The gene expression program elicited by EC extract in 293T cells suggests the establishment of a potential for multilineage differentiation in otherwise more developmentally restricted cells. An indicator of dedifferentiation is the downregulation of genes indicative of a differentiated state. This is exemplified by the downregulation of many 293T cell genes and repression of *LMNA* (see Sect. 5.3.3). *LMNA* downregulation appears specific for extracts of undifferentiated cells, which do not express lamin A/C. In contrast, cardiomyocyte extracts upregulate *LMNA* expression in adipose stem cells, an event that correlates with differentiation toward a cardiomyocyte phenotype (Gaustad et al. 2004), and *LMNA* is reactivated upon retinoic acid-mediated differentiation of EC extract-treated cells (Taranger et al., personal communication). Thus, the transcriptional status of *LMNA* provides a direct assessment of (de)differentiation transitions mediated by cell extracts. The mechanism of gene inactivation mediated by cellular extracts is not known. Nevertheless, evidence for the downregulation of many genes by single small interfering RNAs (Mathieu and Bender 2004), possibly by controlling DNA methylation

(Matzke and Birchler 2005), raises the hypothesis of a contribution of small RNAs in extract-based reprogramming.

Induction of Embryonic and Stem Cell Genes An indicator of induction of pluripotency in EC extract-treated cells is the upregulation of genes characteristic of undifferentiated ECCs or ESCs (Table 1). We found that several embryonic, germ cell, and stem cell genes are activated to levels similar to those of ECCs. The *Oct4* gene is expressed in ESCs to maintain pluripotency, and the Oct4 protein acts by binding to a subset of target genes including *SOX2* (Avilion et al. 2003), *UTF1* (Okuda et al. 1998; Tomioka et al. 2002), and *REX1/DRN3* (Okuda et al. 1998). All these genes were found to be upregulated by the EC extract. Furthermore, because *UTF1* expression requires synergistic activities of Oct4 and Sox2 (Nishimoto et al. 1999), it is likely that Oct4-dependent functions are established in the reprogrammed cells. Telomerase (*TERT*) and telomerase-associated factor 1 (*TERF1*) are also increasingly upregulated. Other pluripotency markers upregulated include placental alkaline phosphatase (*APL1*), leukemia inhibitory factor (*LIF*), stem cell growth factor beta (*SCGF*) and germ cell nuclear factor (*GCNF*). Upregulation of activation of these genes was shown to occur for at least 2 months in culture, but again, how stable these changes are in the long run remains to be determined.

Induction of Multilineage Priming Genes In addition to expressing genes characteristic of "stemness," EC cell extract-treated cells also express genes suggestive of a potential for multiple lineage differentiation (Table 1). These include markers of osteogenic, endothelial, myogenic, neurogenic, adipogenic, and chondrogenic lineages (Boquest et al. 2005). Interestingly, multilineage priming is a hallmark of hematopoietic stem cells (Akashi et al. 2003) and mesenchymal stem cells from bone marrow (Woodbury et al. 2002) and adipose tissue (Boquest et al. 2005). Multilineage priming may reflect the ability of these stem cells to differentiate into a specific cell type in the tissue in which they reside. Expression of multilineage priming genes in EC cell extract-treated cells suggests that the cells may be induced to "survey" their environment and respond to ad hoc stimulation. It seems, therefore, that similarly to so-

Table 1. Some genes characteristic of undifferentiated stem cells and of multi-lineage priming induced in ECC extract-treated 293T cells

Gene name	Description
Embryonic and stem cell genes	
OCT4 and Oct4-responsive genes	
POU5F1	POU domain, class 5, TF1 (Oct4)
SOX2	Sex determining region Y-box 2 (Sox2)
UTF1	Undiff. embryonic cell transcription factor 1
REX1	Deoxyribonuclease III (Drn3)
FOXD3	Forkhead box D3 (FoxD3)
Telomerase and associated factors	
TERT	Telomerase reverse transcriptase
TERF1	Telomerase-associated factor 1
TERF2	Telomerase-associated factor 2
Others	
POU3F1	POU domain, class 3, TF1 (Oct6)
ALP1	Placental alkaline phosphatase 1
CD44	CD44 antigen (homing function)
LIF	Leukemia inhibitory factor
SCGF	Stem cell growth factor beta
GCNF	Germ cell nuclear factor
SPINK2	Serine protease inhibitor, Kazal type, 2
DKK2	Dickkopf (*Xenopus laevis*) homolog 2
Markers of induction of lineage-specific differentiation	
Osteogenic differentiation	
BMP1	Bone morphogenic protein 1
BMP2	Bone morphogenic protein 2
OGN	Osteoglycin
CTSK	Cathepsin K
TNFRSF11B	Osteoprotegerin
Endothelial lineage	
VWF	Von Willebrand Factor
NOS3	Nitric oxide synthase 3 (endothelial cell)
MYF5	Myogenic factor 5
TMP1	Tropomyosin 1 alpha
MYH11	Myosin, heavy polypeptide 11
Neurogenic lineage	
NTS	Neurotensin
NRG1	Neuregulin 1 isoform gamma
MBP	Myelin basic protein
MOBP	Myelin-associated oligodend. basic protein
NCAM1	Neural cell adhesion molecule 1
CD56	Neural cell adhesion molecule CD56

Table 1. (continued)

Gene name	Description
Adipogenic lineage	
APOA2	Apolipoprotein A-II
APOD	Apolipoprotein D
APOE	Apolipoprotein E
APOC1	Apolipoprotein C1
PPARG2	Peroxisome prolif. activated receptor 2
FAD1	Fatty acid desaturase 1
Chondrogenic lineage	
COL4A3	Collagen type IV alpha 3
COL5A2	Collagen, type V, alpha 2
COL8A1	Collagen, type VIII, alpha 1
COL11A1	Collagen, type X1, alpha 1
CSPG2	Chondroitin sulfate proteoglycan 2 (versican)
AGC1	Aggrecan 1
FN1	Fibronectin 1
DSPG3	Dermatan sulphate proteoglycan 3

matic stem cells, the transcriptional signature of EC cell extract-treated cells crosses germ layer boundaries.

5.3.4   How Plastic Are Extract-Treated Cells?

In vitro data suggest that the differentiation plasticity of EC extract-treated cells is enhanced. Exposure of EC extract-treated cells to retinoic acid stimulates the emergence of neuronal progenitor cells. The cells display neurite outgrowths (Fig. 4A), express neuron-specific genes, and label positive for several neuronal markers, while expression of *OCT4* is downregulated. In addition, EC extract-treated cells can be induced to acquire characteristics of adipocytes (Fig. 4B) and osteoblasts (Fig. 4C), as shown by Oil-Red-O and Alizarin red staining, respectively. These observations support the gene expression analyses and argue that the ECC extract can enhance the ability of 293T cells to differentiate into ectoderm and mesoderm lineages. Thus, extracts of undifferentiated ECCs can induce markers of dedifferentiation and signs of differentiation plasticity in an otherwise more developmentally restricted cell type. However, in vivo evidence of pluripotency is still lacking, and it would

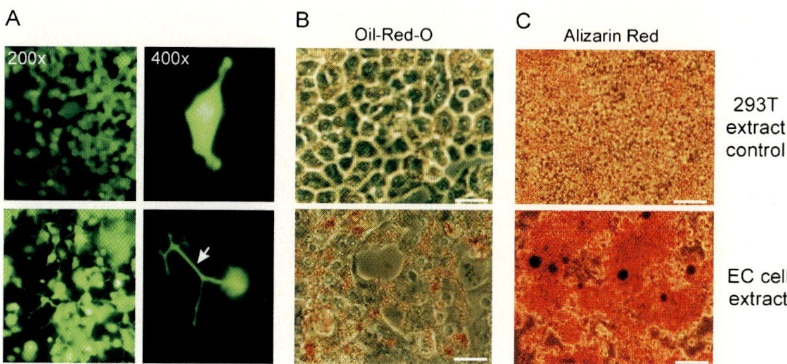

Fig. 4. A–C Induction of neuronal, adipogenic, and osteogenic differentiation of 293T cells reprogrammed with EC extract. **A** Neuronal differentiation, shown here by neurite extensions in genetically labeled EGFP-positive 293T cells (*arrow*). The two sets of panels show GFP fluorescence cells at indicated magnifications. **B** Adipogenic differentiation, as shown by Oil-Red-O staining. **C** Alizarin red staining of mineralized calcium phosphate nodules. Bars, **B**, 20 μm and **C**, 100 μm

be valuable to demonstrate whether reprogrammed cells can contribute to chimerism or form teratomas.

5.4 Chromatin Remodeling Associated with Nuclear Reprogramming

The EC extract retains the ability to elicit epigenetic reprogramming of *OCT4* in 293T cells. Our preliminary data illustrate cytosine demethylation in the *OCT4* promoter (Taranger et al., manuscript in preparation). As mentioned earlier, *Oct4* demethylation in thymocyte nuclei occurs after thymocyte fusion with EGCs or ESCs (Tada et al. 1997, 2001). Furthermore, *Oct4* demethylation is required for *Oct4* transcription after nuclear transplantation into *Xenopus* oocytes (Simonsson and Gurdon 2004); thus our observations are consistent with long-term *OCT4* expression in extract-treated cells. The process driving *OCT4* DNA demethylation remains unclear but seems to require deproteinization (Simonsson and Gurdon 2004) and may involve cleavage of methyl groups (Ramchandani et al. 1999) or cytosine deamination (Morgan et al. 2004). How early Oct4 demethylation occurs after extract exposure is unknown; however, the timing of long-term *Oct4* activation by ESC extract (our unpublished observations) is consistent with the time interval observed between introduction of nuclei into oocytes and *Oct4* demethylation in *Xenopus* oocytes (Simonsson and Gurdon 2004). The ability to induce DNA demethylation in bulk cells or nuclei incubated in extracts raises the possibility of isolating the DNA demethylation activity involved.

Alteration of gene expression in extract-treated cells implies a global and locus-specific remodeling of chromatin. Remodeling of mammalian chromatin by *Xenopus* egg extract depends on ATPase activity of the SWI/SNF chromatin remodeling complex (Kikyo et al. 2000) and in a similar system, the SWI/SNF component BRG1 was shown to be involved in *OCT4* activation by *Xenopus* egg extract (Hansis et al. 2004). Our unpublished observations also support a role of BRG1 in *Oct4* activation by mouse ESC extract in 3T3 fibroblasts. Thus, *OCT4* DNA demethylation and histone H4 hyperacetylation at the *IL2* locus in cells treated with T cell extract (Håkelien et al. 2002) provide evidence that cell extracts can elicit locus-specific chromatin remodeling in exogenous nuclei and cells. Conceivably, controlled manipulations of epigenetic alterations may enhance the heritability of gene expression in reprogrammed cells and may prove beneficial for stably reprogramming cell fate in a therapeutic context.

5.5 Perspectives

5.5.1 Innovative Aspects of Extract-Mediated Cell Reprogramming

If extract-based stem cell differentiation or nuclear reprogramming proved functional on a large scale, the technology could potentially present medical and societal benefits. With the exception of somatic cell nuclear transfer, most current strategies for controlled differentiation are based on a surface or nuclear receptor-mediated approach. Manipulation of cell fate with extract relies on permeabilized cells; thus differentiation or reprogramming molecules access the cellular interior directly.

This may not only render the strategy more effective (Qin et al. 2005), but also has the advantage of being useful without having a great deal of prior knowledge of regulatory mechanisms controlling cell function. Further, although this remains to be extensively tested, in vitro reprogramming may be applied to a variety of cell types including ESCs (Qin et al. 2005), somatic stem cells (Gaustad et al. 2004), and primary cells (Håkelien et al. 2002, 2004; Hansis et al. 2004); thus the technology may enable the production of isogenic cells for regenerative medicine. The target cell type can also be multiple and if truly pluripotent cells could be generated, the technique would clearly present potential for treating many diseases. In addition, unlike reprogramming by nuclear transplantation which utilizes oocytes, pluripotent cultured cells available in large amounts may be used as a source of reprogramming material. Last but not least, ethical and legal issues regarding human cloning and the production of embryonic stem cells from human embryos may be avoided, which would make the approach applicable in a greater number of countries.

Despite promising results, application of an extract-based nuclear reprogramming technology to produce therapeutic replacement cells requires significant development. The approach involves extensive cell manipulation. Thus, phenotype and genotype of the reprogrammed cells need to be extensively characterized to ensure that no genetic perturbations result from the process. Furthermore, target cell phenotypes have been shown to persist for a few months and to our knowledge no data exist to date the long-term stability of the changes elicited. Clearly, efforts need to be put into characterizing and manipulating the epigenetic status of the cells prior to or after exposure to a reprogramming extract. Lastly, it will be important to develop optimal culture systems for the chosen target cell type, or to rapidly transfer the treated cells into the appropriate host environment. Collectively, these components are likely to assist in enhancing the stability of the reprogrammed phenotype.

5.5.2 Cell Extracts as a Tool for Investigating Nuclear Reprogramming Mechanisms

Despite advances in the identification of epigenetic events that accompany reprogramming of specific genes (Simonsson and Gurdon 2004), the efficiency of cell differentiation and nuclear reprogramming into various cell lineages is bound to remain limited without a robust understanding underlying molecular processes. The versatility of cell extracts to elicit markers of transdifferentiation, dedifferentiation, or stem cell differentiation makes them potentially powerful tools for analyzing nuclear reprogramming, at least as it takes place in vitro. Extracts can be manipulated to add or deplete components (Hansis et al. 2004), and fractionation schemes can be designed for identifying molecules central to the establishment of pluripotency or regulation of differentiation. Additionally, because not only cells but also nuclei (Landsverk et al. 2002) or chromatin (Kikyo et al. 2000; Sullivan ct al. 2004) can be treated with extracts, cell-free systems create possibilities for directly manipulating chromatin and the epigenome. It will also be exciting to explore the outcome of similar approaches in other systems such as ESCs or EGs, whether these are used as a source of reprogramming material or are themselves programmed into specific lineages. A clearer understanding of nuclear reprogramming will not only lead to a better appreciation of (de)differentiation, cloning, or stem cell biology, but may also be relevant to the aberrant programming that occurs in life-threatening disorders such as cancer, developmental defects, or premature aging, to name a few.

Acknowledgements. Our work is supported by the Research Council of Norway, the Norwegian Cancer Society, and the University of Oslo.

References

Akashi K, He X, Chen J, Iwasaki H, Niu C, Steenhard B, Zhang J, Haug J, Li L (2003) Transcriptional accessibility for genes of multiple tissues and hematopoietic lineages is hierarchically controlled during early hematopoiesis. Blood 101:383–389

Avilion AA, Nicolis SK, Pevny LH, Perez L, Vivian N, Lovell-Badge R (2003) Multipotent cell lineages in early mouse development depend on SOX2 function. Genes Dev 17:126–140

Blelloch RH, Hochedlinger K, Yamada Y, Brennan C, Kim M, Mintz B, Chin L, Jaenisch R (2004) Nuclear cloning of embryonal carcinoma cells. Proc Natl Acad Sci U S A 101:13985–13990

Boquest AC, Shahdadfar A, Fronsdal K, Sigurjonsson O, Tunheim SH, Collas P, Brinchmann JE (2005) Isolation and transcription profiling of purified uncultured human stromal stem cells: alteration of gene expression following in vitro cell culture. Mol Biol Cell 16:1131–1141

Byrne JA, Simonsson S, Western PS, Gurdon JB (2003) Nuclei of adult mammalian somatic cells are directly reprogrammed to oct-4 stem cell gene expression by amphibian oocytes. Curr Biol 13:1206–1213

Chambers I, Smith A (2004) Self-renewal of teratocarcinoma and embryonic stem cells. Oncogene 23:7150–7160

Cibelli JB, Stice SL, Golueke PJ, Kane JJ, Jerry J, Blackwell C, Ponce DLF, Robl JM (1998) Transgenic bovine chimeric offspring produced from somatic cell-derived stem-like cells. Nat Biotechnol 16:642–646

Collas P, Håkelien AM (2003) Teaching cells new tricks. Trends Biotechnol 21:354–361

Flasza M, Shering AF, Smith K, Andrews PW, Talley P, Johnson PA (2003) Reprogramming in inter-species embryonal carcinoma-somatic cell hybrids induces expression of pluripotency and differentiation markers. Cloning Stem Cells 5:339–354

Gaustad KG, Boquest AC, Anderson BE, Gerdes AM, Collas P (2004) Differentiation of human adipose tissue stem cells using extracts of rat cardiomyocytes. Biochem Biophys Res Commun 314:420–427

Gurdon JB, Byrne JA (2003) The first half-century of nuclear transplantation. Proc Natl Acad Sci U S A 100:8048–8052

Håkelien AM, Landsverk HB, Robl JM, Skålhegg BS, Collas P (2002) Reprogramming fibroblasts to express T-cell functions using cell extracts. Nat Biotechnol 20:460–466

Håkelien AM, Gaustad KG, Collas P (2004) Transient alteration of cell fate using a nuclear and cytoplasmic extract of an insulinoma cell line. Biochem Biophys Res Commun 316:834–841

Håkelien AM, Gaustad KG, Taranger CK, Skalhegg BS, Kuntziger T, Collas P (2005) Long-term in vitro, cell-type-specific genome-wide reprogramming of gene expression. Exp Cell Res 309:32–47

Hansis C, Barreto G, Maltry N, Niehrs C (2004) Nuclear reprogramming of human somatic cells by xenopus egg extract requires BRG1. Curr Biol 14:1475–1480

Hutchison CJ, Worman HJ (2004) A-type lamins: guardians of the soma? Nat Cell Biol 6:1062–1067

Kikyo N, Wade PA, Guschin D, Ge H, Wolffe AP (2000) Active remodeling of somatic nuclei in egg cytoplasm by the nucleosomal ATPase ISWI. Science 289:2360–2362

Klevecz RR, Bolen J, Forrest G, Murray DB (2004) A genomewide oscillation in transcription gates DNA replication and cell cycle. Proc Natl Acad Sci U S A 101:1200–1205

Landsverk HB, Håkelien AM, Küntziger T, Robl JM, Skålhegg BS, Collas P (2002) Reprogrammed gene expression in a somatic cell-free extract. EMBO Rep 3:384–389

Mathieu O, Bender J (2004) RNA-directed DNA methylation. J Cell Sci 117:4881–4888

Matzke MA, Birchler JA (2005) RNAi-mediated pathways in the nucleus. Nat Rev Genet 6:24–35

McGann CJ, Odelberg SJ, Keating MT (2001) Mammalian myotube dedifferentiation induced by newt regeneration extract. Proc Natl Acad Sci U S A 98:13699–13704

Miklos GL, Maleszka R (2004) Microarray reality checks in the context of a complex disease. Nat Biotechnol 22:615–621

Morgan HD, Dean W, Coker HA, Reik W, Petersen-Mahrt SK (2004) Activation-induced cytidine deaminase deaminates 5-methylcytosine in DNA and is expressed in pluripotent tissues: implications for epigenetic reprogramming. J Biol Chem 279:52353–52360

Munsie MJ, Michalska AE, O'Brien CM, Trounson AO, Pera MF, Mountford PS (2000) Isolation of pluripotent embryonic stem cells from reprogrammed adult mouse somatic cell nuclei. Curr Biol 10:989–992

Nishimoto M, Fukushima A, Okuda A, Muramatsu M (1999) The gene for the embryonic stem cell coactivator UTF1 carries a regulatory element which selectively interacts with a complex composed of Oct-3/4 and Sox-2. Mol Cell Biol 19:5453–5465

Okuda A, Fukushima A, Nishimoto M, Orimo A, Yamagishi T, Nabeshima Y, Kuro-o M, Nabeshima Y, Boon K, Keaveney M, Stunnenberg HG, Muramatsu M (1998) UTF1, a novel transcriptional coactivator expressed in pluripotent embryonic stem cells and extra-embryonic cells. EMBO J 17:2019–2032

Paulsson J (2004) Summing up the noise in gene networks. Nature 427:415–418

Pells S, Di Domenico AI, Callagher EJ, McWhir J (2002) Multipotentiality of neuronal cells after spontaneous fusion with embryonic stem cells and nuclear reprogramming in vitro. Cloning Stem Cells 4:331–338

Qin M, Tai G, Collas P, Polak JM, Bishop AE (2005) Cell extract-derived differentiation of embryonic stem cells. Stem Cells 23:712–718

Ramchandani S, Bhattacharya SK, Cervoni N, Szyf M (1999) DNA methylation is a reversible biological signal. Proc Natl Acad Sci U S A 96:6107–6112

Simonsson S, Gurdon J (2004) DNA demethylation is necessary for the epigenetic reprogramming of somatic cell nuclei. Nat Cell Biol 6:984–990

Sullivan EJ, Kasinathan S, Kasinathan P, Robl JM, Collas P (2004) Cloned calves from chromatin remodeled in vitro. Biol Reprod 70:146–153

Tada M, Tada T, Lefebvre L, Barton SC, Surani MA (1997) Embryonic germ cells induce epigenetic reprogramming of somatic nucleus in hybrid cells. EMBO J 16:6510–6520

Tada M, Takahama Y, Abe K, Nakastuji N, Tada T (2001) Nuclear reprogramming of somatic cells by in vitro hybridization with ES cells. Curr Biol 11:1553–1558

Terada N, Hamazaki T, Oka M, Hoki M, Mastalerz DM, Nakano Y, Meyer EM, Morel L, Petersen BE, Scott EW (2002) Bone marrow cells adopt the phenotype of other cells by spontaneous cell fusion. Nature 416:542–545

Tomioka M, Nishimoto M, Miyagi S, Katayanagi T, Fukui N, Niwa H, Muramatsu M, Okuda A (2002) Identification of Sox-2 regulatory region which is under the control of Oct-3/4-Sox-2 complex. Nucleic Acids Res 30:3202–3213

Wakayama T, Tabar V, Rodriguez I, Perry AC, Studer L, Mombaerts P (2001) Differentiation of embryonic stem cell lines generated from adult somatic cells by nuclear transfer. Science 292:740–743

Wilmut I, Beaujean N, De Sousa PA, Dinnyes A, King TJ, Paterson LA, Wells DN, Young LE (2002) Somatic cell nuclear transfer. Nature 419:583–586

Woodbury D, Reynolds K, Black IB (2002) Adult bone marrow stromal stem cells express germline, ectodermal, endodermal, and mesodermal genes prior to neurogenesis. J Neurosci Res 69:908–917

Ying QL, Nichols J, Evans EP, Smith AG (2002) Changing potency by spontaneous fusion. Nature 416:545–548

6 Molecular Switches and Developmental Potential of Adult Stem Cells

M. Zenke, T. Hieronymus

6.1   The Development of Blood Cells from Multipotent Hematopoietic Stem Cells 70

6.2   The Flt3$^+$CD11b$^+$ Multipotent Progenitor 71

6.3   The HLH Transcription Factor Id2 in Lineage Choice 72

References 76

Abstract. Stem cell commitment and differentiation entails the successive loss of self-renewal and developmental potential, and results in the final restriction to a terminally differentiated mature cell type. Hematopoiesis, the development of blood cells from hematopoietic stem cells in bone marrow, is particularly well studied, and at different branching points within the hematopoietic system multiple developmental intermediates have been identified. Here we describe a Flt3$^+$CD11b$^+$ multipotent progenitor that can be amplified in vitro by a specific cytokine combination to high cell numbers, and following adoptive transfer into syngeneic mice, it generates dendritic cells but also additional mature cell types. By employing gene expression profiling with DNA microarrays and knockout mouse models, we demonstrate that the helix-loop-helix (HLH) transcription factor Id2 (inhibitor of DNA binding/differentiation 2) acts as a molecular switch in development of Langerhans cells (LCs), the cutaneous contingent of dendritic cells (DCs), and of specific DC subsets and B cells.

6.1 The Development of Blood Cells from Multipotent Hematopoietic Stem Cells

The developmental options available to stem cells involve the successive loss of self-renewal potency and multilineage differentiation potential, and yields fully mature and functional cells. Hematopoiesis represents the development of all mature cells in blood from a population of multipotent hematopoietic stem cells (HSCs), which due to their sustained self-renewal capacity and multilineage differentiation potential maintain blood cell numbers throughout life. To date this adult stem cell differentiation system has been particularly well studied.

Stem cells and multipotent progenitor cells exhibit a broad gene expression repertoire and promiscuously express genes of several lineages but at low levels (Graf 2002; Orkin 2000; Weissman 2000, and references therein). During development of specific hematopoietic lineages, the gene expression repertoire and the options of stem cells and multipotent progenitors become increasingly restricted, leading to the establishment to a distinct lineage from the choice of several. At different branching points within the hematopoietic system, multiple developmental intermediates thus exist that exhibit a progressively restricted expansion capacity and developmental potential.

Among these more restricted progenitor cells are the well-studied common lymphoid progenitor (CLP) and common myeloid progenitor (CMP; Graf 2002; Orkin 2000; Weissman 2000). CLPs can develop into B cells, T cells, NK cells and lymphoid-derived DCs, but not myeloid cells; CMPs develop into all myeloid cells, including macrophages, granulocytes, and myeloid-derived DCs. Recent studies challenge this currently prevailing model of a strict separation into CLP and CMP as the first lineage commitment step in adult hematopoiesis (Adolfsson et al. 2005). It is rather suggested that the first lineage restriction site is the separation of cells with erythroid/megakaryocytic potential from cells with a lymphoid-myeloid potential. Thus, it remains open whether further, thus far unrecognized restriction sites exist that determine hematopoietic cell fate.

6.2 The Flt3$^+$CD11b$^+$ Multipotent Progenitor

Transcriptional profiling by DNA microarrays represents a particularly versatile approach for determining the gene expression repertoire of a given cell population and how different cell types, including stem cells, relate to each other. An appealing question is whether the functional similarities of different stem cell types, such as HSCs, neural stem cells (NSCs) and embryonic stem cells (ESCs), can be defined by the expression of a limited number of "stemness" genes. Surprisingly, such studies have been met only with limited success, identifying only one gene, CD49f/integrin α6, as being expressed in all stem cell types studied (Fortunel et al. 2003; Ivanova et al. 2002; Ramalho-Santos et al. 2002; RD Kirsch, T Hieronymus and M Zenke, unpublished data). However, this observation is interesting and suggests an important function of stem cell interactions with the surrounding tissue, frequently referred to as the stem cell niche. The impact of the stem cell niche on stem cell integrity and maintenance of stem cell numbers is now being intensively studied (Fuchs et al. 2004; Nishimura et al. 2002).

HSCs are frequently obtained according to their specific surface phenotype by immunomagnetic bead selection and/or FACS sorting. By using an alternative approach, we recently identified a multipotent progenitor from mouse bone marrow that expresses various stem cell markers, including CD49f/integrin α6, and also the integrin CD11b/Mac-1, in the following referred to as Flt3$^+$CD11b$^+$ multipotent progenitor (MPP; Hieronymus et al. 2005). Flt3$^+$CD11b$^+$ MPPs can be amplified in vitro by a specific cytokine combination (SCF+Flt3L+IGF-I+hyper-IL-6) to high cell numbers, and following adoptive transfer into syngeneic mice it can generate dendritic cells (DCs; Hieronymus et al. 2005) but also additional mature cell types, indicating a multilineage differentiation potential (T. Hieronymus and M. Zenke, unpublished observations).

An obvious question is how the Flt3$^+$CD11b$^+$ MPP relates to the Flt3$^+$ MPP, CMP, and CLP described in the literature (Akashi et al. 2003; D'Amico and Wu 2003; Graf 2002; Orkin 2000; Weissman 2000). However, all routinely employed protocols for purification of HSCs, MPPs, CLPs, and CMPs from mouse bone marrow use CD11b/Mac-1 antibody in the lineage depletion antibody cocktail to remove macrophages and thus CD11b$^+$ cells are excluded from further analysis.

By employing transcriptional profiling with DNA microarrays and immunophenotyping with specific antibodies, we demonstrated that Flt3$^+$CD11b$^+$ MPPs express several stem cell antigens such as CD93 (Ly68/AA4.1), CD133 (AC133/Prominin), and integrin α6 (CD49f; Hieronymus et al. 2005). Furthermore, the surface marker make-up of Flt3$^+$CD11b$^+$ MPPs was compared to that of highly purified lin$^-$c-kit$^+$Sca1$^+$ HSCs. Flt3$^+$CD11b$^+$ MPPs express 73 of the 206 CD molecules analyzed, while HSCs expressed 59. Surprisingly, the majority of surface antigens (49) were found in both HSCs and Flt3$^+$CD11b$^+$ MPPs, while only 10 and 24 were solely found in HSCs and Flt3$^+$CD11b$^+$ MPPs, respectively (Hieronymus et al. 2005). Flt3$^+$CD11b$^+$ MPPs also express several cytokine receptors that are related to the stem cell phenotype, such as receptors for flt3 ligand (FL) (Flt3/CD135), stem cell factor (SCF; c-kit/CD117), and stromal-cell-derived factor-1 (SDF-1; CXCR4/CD184), and also the IL-6 signal transducer gp130/CD130.

Thus, the Flt3$^+$CD11b$^+$ MPP identified has so far escaped attention since it is depleted in the frequently used stem purification protocols, and therefore future studies aim at determining the molecular and functional properties of this stem cell population.

6.3 The HLH Transcription Factor Id2 in Lineage Choice

Stem/progenitor cell renewal and cell fate choice in hematopoiesis are regulated by various signaling pathways, including cytokines and cytokine receptors, Wnt and Notch factors (Allman et al. 2002; Graf 2002; Lyman and Jacobson 1998; Reya et al. 2003; Weissman 2000). These signaling pathways modulate gene activity of lineage determining transcription factors. Transcription factors were demonstrated to play a central role in lineage commitment and differentiation, and factors that control the development of red blood cells, macrophages, B cells, and T cells have been particularly well studied (Graf 2002; Orkin 2000; Siewecke et al 1998). In contrast, our understanding of the developmental pathways leading to DCs is more limited. In vitro culture systems and transcriptional profiling with DNA microarrays and the use of the knockout mouse model have now provided novel insights into how stem cells are committed and develop into DCs.

DCs are professional antigen-presenting cells and exhibit a pivotal role in antigen-specific immune responses and have been implicated in determining the balance between immunity and maintenance of immunological tolerance (Banchereau and Steinman 1998; Banchereau et al. 2000). DCs occur throughout the organism in both lymphoid and nonlymphoid tissues and are specialized for uptake, transport, processing, and presentation of antigens. DCs constitute a complex system of heterogeneous cells and different DC subsets have been described that differ in tissue distribution, surface phenotype, and specific functions (Banchereau and Steinman 1998; Banchereau et al. 2000). It is now assumed that different DC subsets are derived from different hematopoietic lineages (Ardavin et al. 2001; Liu et al. 2001; Shortman and Liu 2002). However, their lineage origin and the underlying molecular mechanisms that determine DC development have remained largely unclear or highly controversial (Ardavin 2003; Shortman and Liu 2002).

Over the past several years, we have developed culture systems that recapitulate DC development from HSCs in vitro under well-defined conditions (Hacker et al. 2003; Hieronymus et al. 2005; Ju et al 2003). In these two-step culture systems, HSCs are isolated from mouse bone marrow or human cord blood, and cells are grown with specific growth factors and cytokines to obtain homogenous populations of stem/progenitor cells. To induce DC differentiation, cells are treated with differentiation factors, such as GM-CSF, and cells undergo cell cycle arrest and differentiate into fully mature and functional DCs.

During hematopoiesis, lineage specification and development of mature blood cells involve the activation of lineage-specific genes and the selective repression of genes for alternative lineages, thereby leading to the establishment of a lineage-specific differentiation program. Gene expression profiling with DNA microarrays makes it possible to determine the gene repertoire of a given cell population on a genome-wide scale and to monitor the ongoing changes in gene expression when cells differentiate (Ju and Zenke 2004). By employing this approach, we identified the helix-loop-helix (HLH) transcription factor Id2 (inhibitor of DNA binding/differentiation 2) as a determining factor for development of Langerhans cells (LCs), the cutaneous contingent of DCs, and of specific DC subsets and B cells (Hacker et al. 2003; Ju and Zenke 2004).

HLH transcription factors have been shown to be important for cell fate decisions in a number of systems, including hematopoietic cells (Massari and Murre 2000; Yokota 2001). Activating HLH factors function as transcriptional activators and are antagonized by yet another class of HLH factors that act as transcriptional repressors, referred to as Id proteins (inhibitors of DNA binding/differentiation; Massari and Murre 2000; Yokota 2001). Id proteins contain the highly conserved HLH domain but lack the adjacent basic region that is required for DNA binding. Thus, heterodimerization of Id proteins with HLH activators results in a complex that is incapable of DNA binding. The function of Id proteins has recently been addressed in gene inactivation studies in mice. $Id1^{-/-}$ mice do not exhibit overt abnormalities. $Id2^{-/-}$ mice are compromised in the development of the natural killer (NK) cell, the mammary gland and the nasopharyngeal-associated lymphoid tissue (NALT), lack lymph nodes and Peyer's patches, and – as demonstrated by our work – specific subsets of DCs (Hacker et al. 2003; Ju and Zenke 2004; Kusonoki et al. 2003; Yokota 2001; Yokota et al. 1999). The analysis of $Id3^{-/-}$ mice suggested that Id3 is an important component of TGFβ1 signaling in B cells (Kee et al. 2001).

During DC development from stem cells, Id2 RNA and protein were found to be effectively upregulated. To determine the impact of Id2 for DC development in vivo, $Id2^{-/-}$ mice were studied (Hacker et al. 2003; Kusonoki et al. 2003; Yokota et al. 1999). It was found that $Id2^{-/-}$ mice lack LCs and the splenic $CD8\alpha^+$ DC subset (Fig. 1). The number of $CD11c^+CD11b^+$ DCs, the classical tissue DCs that are found in almost all tissues, was unchanged and the frequency of $CD11c^+CD11b^-$ plasmacytoid DCs (pDCs) was elevated (Hacker et al. 2003). The number of B cells in $Id2^{-/-}$ was increased, as demonstrated by Becker-Herman et al. (2002). Thus, Id2 has a selective impact on the development of specific DC subsets and on B cells.

$TGF\beta1^{-/-}$ mice also lack LCs and we found that TGFβ1 induces Id2 expression (Hacker et al. 2003). This induction of Id2 by TGFβ1 also occurred in the presence of the protein biosynthesis inhibitor cycloheximide, indicating that TGFβ1 signaling – most probably via the Smad signaling pathway – has a direct effect on Id2 transcription (Hacker et al. 2003; XS Ju and M Zenke, unpublished observation). We also demonstrated by bone marrow transfer experiments that the effect of Id2 on lin-

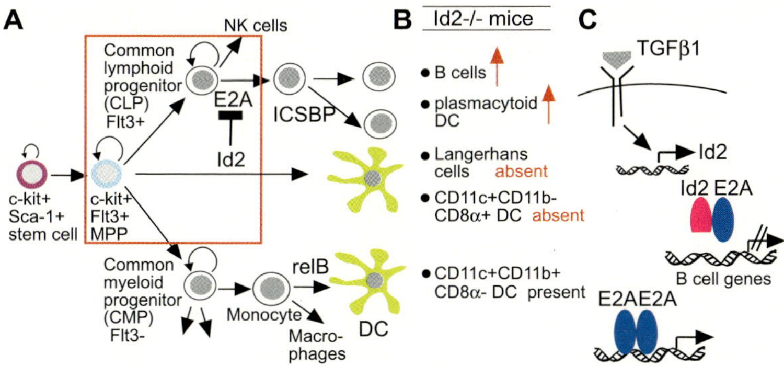

Fig. 1. A–C The TGFβ1/Id2 signaling pathway in stem/progenitor cell commitment and differentiation into DCs. **A** Lin-c$^-$kit$^+$Sca-1$^+$ hematopoietic stem cells, c-kit$^+$Flt3$^+$ multipotent progenitors (*MPP*), common myeloid progenitor (*CMP*), and common lymphoid progenitor (*CLP*) and various mature cell types, and their developmental relationship are indicated. *DC*, dendritic cells; *NK*, natural killer cells. The positions where the transcription factors Id2, E2A, relB, and ICSBP act are indicated. **B** Phenotype of Id2$^{-/-}$ mice. **C** The TGFβ1/Id2 signaling pathway

eage choice and DC development is cell autonomous. Finally, to obtain insight into the genes that are regulated by Id2, CD11c$^+$CD11b$^+$ DCs from Id2$^{+/+}$ and Id2$^{-/-}$ mice were isolated and analyzed by gene expression profiling with DNA microarrays. It was found that Id2 represses B cell genes in DCs (Hacker et al. 2003). This study identified a novel TGFβ1/Id2 signaling pathway in stem/progenitor cell commitment and DC development, and suggests that Id2 represents a key component in lineage choice of B cells and DCs.

Therefore a model presents itself where Id2 represents a factor that – by silencing the activity of the activating HLH factor such as E2A – regulates the propensity of a stem cell and/or multipotent progenitor cell to undergo DC development at the expense of B cell development: low or no expression of Id2 and high expression of E2A supports B cell development, whereas upregulation of Id2 (either developmentally controlled or following induction by TGFβ1) blunts E2A activity and B cell development and thus favors differentiation into DCs (Hacker et al. 2003; Ju and Zenke 2004).

While Id2$^{-/-}$ mice lack LCs and splenic CD8α^+ DCs, and display increased frequencies of pDCs and B cells, other knockout mice lack other or identical DC subsets: RelB$^{-/-}$ mice lack the classical CD11c$^+$CD11b$^+$ DCs (Burkley et al. 1995; Weih et al. 1995; Wu et al. 1998) and interferon consensus sequence-binding protein (ICSBP)/IRF-8$^{-/-}$ mice lack pDCs and LCs (Alberti et al. 2003; Schiavoni et al. 2002; Schiavoni et al. 2004; Tamura et al. 2005; Tsujimura et al. 2003). Mice deficient in IRF-2 (interferon regulatory factor-2), IRF-4, and Runx transcription factors also exhibit deficiencies in the DC compartment (Ardavin 2003; Ichikawa et al. 2004; Fainaru et al. 2004; Suzuki et al. 2004; Tamura et al. 2005). Thus, these studies now provide information about a "road map" of how HSCs and multipotent progenitors develop – via the myeloid and/or lymphoid pathway – to give rise to DCs (Fig. 1).

Acknowledgements. We would like to thank our colleagues for discussion and A. Offergeld for secretarial assistance. The work described was funded by grants of the German Research Foundation DFG Ze432/1–3 and Ze432/2–3 and by the BMBF, Bonn, Germany.

References

Adolfsson J, Mansson R, Buza-Vidas N, Hultquist A, Liuba K, Jensen CT, Bryder D, Yang L, Borge OJ, Thoren LA, Anderson K, Sitnicka E, Sasaki Y, Sigvardsson M, Jacobsen SE (2005) Identification of Flt3+ lympho-myeloid stem cells lacking erythro-megakaryocytic potential: a revised road map for adult blood lineage commitment. Cell 121:295–306

Akashi K, He X, Chen J, Iwasaki H, Niu C, Steenhard B, Zhang J, Haug J, Li L (2003) Transcriptional accessibility for genes of multiple tissues and hematopoietic lineages is hierarchically controlled during early hematopoiesis. Blood 101:383–389

Aliberti J, Schulz O, Pennington DJ, Tsujimura H, Reis e Sousa C, Ozato K, Sher A (2003) Essential role for ICSBP in the in vivo development of murine CD8alpha+ dendritic cells. Blood 101:305–310

Allman D, Aster JC, Pear WS (2002) Notch signaling in hematopoiesis and early lymphocyte development. Immunol Rev 187:75–86

Ardavin C (2003) Origin, precursors and differentiation of mouse dendritic cells. Nat Rev Immunol 3:582–590

Banchereau J, Steinman RM (1998) Dendritic cells and the control of immunity. Nature 392:245–252

Banchereau J, Briere F, Caux C, Davoust J, Lebecque S, Liu YJ, Pulendran B, Palucka K (2000) Immunobiology of dendritic cells. Annu Rev Immunol 18:767–811

Becker-Herman S, Lantner F, Shachar I (2002) Id2 negatively regulates B cell differentiation in the spleen. J Immunol 168:5507–5513

Burkly L, Hession C, Ogata L, Reilly C, Marconi LA, Olson D, Tizard R, Cate R, Lo D (1995) Expression of relB is required for the development of thymic medulla and dendritic cells. Nature 373:531–536

D'Amico A, Wu L (2003) The early progenitors of mouse dendritic cells and plasmacytoid predendritic cells are within the bone marrow hemopoietic precursors expressing Flt3. J Exp Med 198:293–303

Fainaru O, Woolf E, Lotem J, Yarmus M, Brenner O, Goldenberg D, Negreanu V, Bernstein Y, Levanon D, Jung S, Groner Y (2004) Runx3 regulates mouse TGF-beta-mediated dendritic cell function and its absence results in airway inflammation. EMBO J 23:969–979

Fortunel NO, Otu HH, Ng HH, Chen J, Mu X, Chevassut T, Li X, Joseph M, Bailey C, Hatzfeld JA, Hatzfeld A, Usta F, Vega VB, Long PM, Libermann TA, Lim B (2003) Comment on " 'Stemness': transcriptional profiling of embryonic and adult stem cells" and "a stem cell molecular signature". Science 302:393b

Fuchs E, Tumbar T, Guasch G (2004) Socializing with the neighbors: stem cells and their niche. Cell 116:769–778

Graf T (2002) Differentiation plasticity of hematopoietic cells. Blood 99:3089–3101

Hacker C, Kirsch RD, Ju XS, Hieronymus T, Gust TC, Kuhl C, Jorgas T, Kurz SM, Rose-John S, Yokota Y, Zenke M (2003) Transcriptional profiling identifies Id2 function in dendritic cell development. Nat Immunol 4:380–386

Hieronymus T, Gust TC, Kirsch RD, Jorgas T, Blendinger G, Goncharenko M, Supplitt K, Rose-John S, Muller AM, Zenke M (2005) Progressive and controlled development of mouse dendritic cells from Flt3+CD11b+ progenitors in vitro. J Immunol 174:2552–2562

Ichikawa E, Hida S, Omatsu Y, Shimoyama S, Takahara K, Miyagawa S, Inaba K, Taki S (2004) Defective development of splenic and epidermal CD4+ dendritic cells in mice deficient for IFN regulatory factor-2. Proc Natl Acad Sci U S A 101:3909–3914

Ivanova NB, Dimos JT, Schaniel C, Hackney JA, Moore KA, Lemischka IR (2002) A stem cell molecular signature. Science 298:601–604

Ju X-S, Zenke M (2004) Gene expression profiling of dendritic cells by DNA microarrays. Immunobiology 209:155–161

Ju XS, Hacker C, Madruga J, Kurz SM, Knespel S, Blendinger G, Rose-John S, Martin Z (2003) Towards determining the differentiation program of antigen-presenting dendritic cells by transcriptional profiling. Eur J Cell Biol 82:75–86

Kee BL, Rivera RR, Murre C (2001) Id3 inhibits B lymphocyte progenitor growth and survival in response to TGF-beta. Nat Immunol 2:242–247

Kusunoki T, Sugai M, Katakai T, Omatsu Y, Iyoda T, Inaba K, Nakahata T, Shimizu A, Yokota Y (2003) TH2 dominance and defective development of a CD8+ dendritic cell subset in Id2-deficient mice. J Allergy Clin Immunol 111:136–142

Liu YJ, Kanzler H, Soumelis V, Gilliet M (2001) Dendritic cell lineage, plasticity and cross-regulation. Nat Immunol 2:585–589

Lyman SD, Jacobsen SE (1998) C-kit ligand and Flt3 ligand: stem/progenitor cell factors with overlapping yet distinct activities. Blood 91:1101–1134

Massari ME, Murre C (2000) Helix-loop-helix proteins: regulators of transcription in eucaryotic organisms. Mol Cell Biol 20:429–440

Nishimura EK, Jordan SA, Oshima H, Yoshida H, Osawa M, Moriyama M, Jackson IJ, Barrandon Y, Miyachi Y, Nishikawa S (2002) Dominant role of the niche in melanocyte stem-cell fate determination. Nature 416:854–860

Orkin SH (2000) Diversification of haematopoietic stem cells to specific lineages. Nat Rev Genet 1:57–64

Ramalho-Santos M, Yoon S, Matsuzaki Y, Mulligan RC, Melton DA (2002) "Stemness": transcriptional profiling of embryonic and adult stem cells. Science 298:597–600

Reya T, Duncan AW, Ailles L, Domen J, Scherer DC, Willert K, Hintz L, Nusse R, Weissman IL (2003) A role for Wnt signalling in self-renewal of haematopoietic stem cells. Nature 423:409–414

Schiavoni G, Mattei F, Sestili P, Borghi P, Venditti M, Morse HC 3rd, Belardelli F, Gabriele L (2002) ICSBP is essential for the development of mouse type I interferon-producing cells and for the generation and activation of CD8alpha(+) dendritic cells. J Exp Med 196:1415–1425

Schiavoni G, Mattei F, Borghi P, Sestili P, Venditti M, Morse HC 3rd, Belardelli F, Gabriele L (2004) ICSBP is critically involved in the normal development and trafficking of Langerhans cells and dermal dendritic cells. Blood 103:2221–2228

Shortman K, Liu YJ (2002) Mouse and human dendritic cell subtypes. Nature Rev Immunol 2:151–161

Sieweke MH, Graf T (1998) A transcription factor party during blood cell differentiation. Curr Opin Genet Dev 8:545–551

Suzuki S, Honma K, Matsuyama T, Suzuki K, Toriyama K, Akitoyo I, Yamamoto K, Suematsu T, Nakamura M, Yui K, Kumatori A (2004) Critical roles of interferon regulatory factor 4 in CD11bhighCD8alpha-dendritic cell development. Proc Natl Acad Sci U S A 101:8981–8986

Tamura T, Nagamura-Inoue T, Shmeltzer Z, Kuwata T, Ozato K (2000) ICSBP directs bipotential myeloid progenitor cells to differentiate into mature macrophages. Immunity 13:155–165

Tamura T, Tailor P, Yamaoka K, Kong HJ, Tsujimura H, O'Shea JJ, Singh H, Ozato K (2005) IFN regulatory factor-4 and -8 govern dendritic cell subset development and their functional diversity. J Immunol 174:2573–2581

Tsujimura H, Tamura T, Ozato K (2003) Cutting edge: IFN consensus sequence binding protein/IFN regulatory factor 8 drives the development of type I IFN-producing plasmacytoid dendritic cells. J Immunol 170:1131–1135

Weih F, Carrasco D, Durham SK, Barton DS, Rizzo CA, Ryseck RP, Lira SA, Bravo R (1995) Multiorgan inflammation and hematopoietic abnormalities in mice with a targeted disruption of RelB, a member of the NF-kappa B/Rel family. Cell 80:331–340

Weissman IL (2000) Stem cells: units of development, units of regeneration, and units in evolution. Cell 100:157–168

Wu L, D'Amico A, Winkel KD, Suter M, Lo D, Shortman K (1998) RelB is essential for the development of myeloid-related CD8alpha– dendritic cells but not of lymphoid-related CD8alpha+ dendritic cells. Immunity 9:839–847

Yokota Y (2001) Id and development. Oncogene 20:8290–8298

Yokota Y, Mansouri A, Mori S, Sugawara S, Adachi S, Nishikawa S, Gruss P (1999) Development of peripheral lymphoid organs and natural killer cells depends on the helix-loop-helix inhibitor Id2. Nature 397:702–706

7 Adult Small Intestinal Stem Cells: Identification, Location, Characteristics, and Clinical Applications

C.S. Potten, J.R. Ellis

References 97

Abstract. There are few systems which enable adult tissue stem cells to be studied. However, the gastrointestinal tract with its high degree of polarity, well-defined cell migratory pathways, and dynamic cell replacement is a model tissue providing unique opportunities for stem cell study. Lineage tracking indicates that all cell replacement originates at well-defined stem cell positions, with an associated slower cell cycle. Radiobiological studies suggest a hierarchical stem cell compartment (actual and potential stem cells). Actual stem cells have an intolerance of genotoxic damage and die via apoptosis. Stem cells also selectively sort the old and new DNA strands at division, retaining the replication error free strands in the stem cell daughter. High genotoxic sensitivity and selective sorting of old and new DNA strands, provides extremely effective protective mechanisms against both replication and random errors. This provides a new explanation for the low cancer risk in the small intestine.

The small intestine, like several other adult tissues, is structurally and functionally organised into discrete units of proliferation, the crypts, and units of functionally differentiated cells, the villi. The details of the histological structure and basic cell kinetics of the various regions of the small and large bowel have been extensively studied for more than 50 years and widely reviewed (Potten 1995, 1998; Potten et al. 1997; Marshman et al. 2002). During the last 25 years the concept of adult tissue stem cells has evolved and developed into a major area of interest. It is now generally accepted that each of the small intestinal crypts in the mouse, which contain about 250 cells in total, has approximately five cell lineages with each lineage derived from a self-maintaining lineage ancestor stem cell. In the murine small intestine, there are about six generations of dividing transit cells within each lineage. The central region of the large bowel in the mouse has a similar number of lineages but each lineage may have between six and eight cell generations. Similar lineage organisation can be inferred for the human small and large bowel but the number of lineages per crypt, and hence lineage ancestor stem cells, remain somewhat unclear.

Alternative models of the crypt organisation, notably one consisting of a single ultimate stem cell with a consequent longer lineage, cannot be completely ruled out, but the consensus view and the bulk of available data suggest a model of between four and six lineages per crypt. Other even more complex models involving master crypts with super stem cells from which other crypts and their stem cells are derived can also be envisaged, but there is little firm evidence to support such a hypothesis.

So the indications are that the crypt contains between four and six lineage ancestor stem cells (for convenience we will call this five). This number is supported by evidence from estimates of the number of clonogenic stem cells at low doses of radiation, mathematical modelling of the cell kinetics, the number of label-retaining cells and the number of cells that are apparently selectively segregating the template strands of DNA, and finally, the number of apoptosis-susceptible cells. Other data using Msi-1 expression, intense P53 expression and early bromodeoxyuridine (BrdU) data for the regeneration response also suggest a small number of important cells at cell positions 4–5.

Slight differences exist between the size of the differentiated and proliferative units as one moves along the small intestine and into the large intestine. In the terminal ileum each crypt may provide cells for more than one villus and each villus is served by between six and ten crypts (see Fig. 1). A variety of old and modern lineage tracking studies

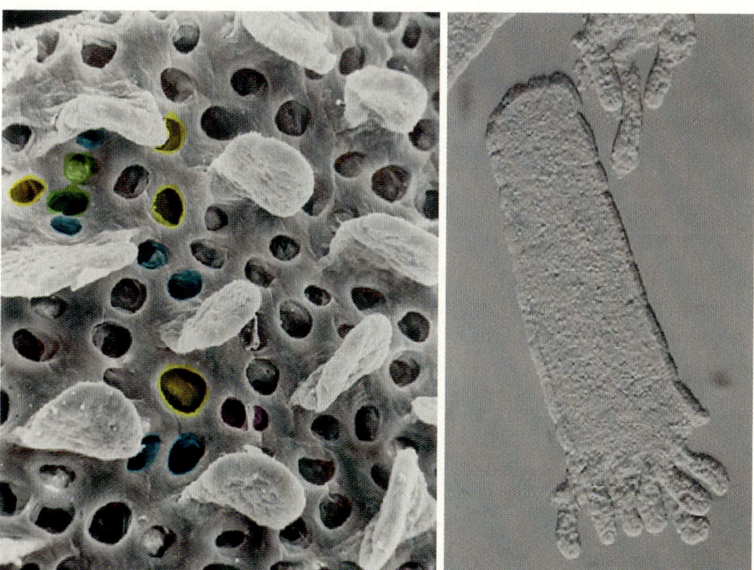

Fig. 1. The 3D relationship between crypts and villi in mouse small intestine. *Left panel*: Scanning electron microscope image of the tissue after the epithelium has been removed. Various spatial relationships can be seen for the crypt to villus interaction. Some crypts (*yellow*) can clearly serve more than one villus. Others (*blue*) appear to be more unidirectional. *Right panel*: A dissected whole villus showing the cluster of associated crypts

have been performed on this dynamic process (Qiu et al. 1994; Winton et al. 1998; Winton and Ponder 1990; Wong et al. 2000). Such studies have enabled the migratory pathways to be mapped, the velocity of the cell movement to be determined and the stem cell lineages to be identified. Cells move as vertical columns, at least once on the villus, where they move with a velocity of between one and two cell diameters per hour (Kaur and Potten 1986). As a consequence of this migratory activity, new cells born in the upper region of the proliferative zone of the crypt of the mouse reach the tip of the villus in about 3 days, which represents their total life expectancy in the tissue. Cells born from lower in the crypt may have a life expectancy of about 5 days. Thus within 3–

5 days of birth from mitosis, most cells have completed their functional life span and are shed into the lumen of the intestine as senescent or apoptotic cells.

A vast amount of experimental data have been accumulated relating to the cell replacement processes in this tissue, and the quantity of data has enabled sophisticated, mathematical modelling and computer simulation studies to be undertaken (Potten and Loeffler 1990; Meineke et al. 2001).

The cell migration studies where the cell velocity was determined at each cell position along the long axis of the crypt provided clear evidence that all cells in the epithelium could ultimately be traced to a very specific position in the crypt in relation to its long axis (Kaur and Potten 1986; Qiu et al. 1994). The conclusions from these studies were that the origin of the lineages, and therefore the location of the stem cells, could be traced back to the fourth or fifth cell position from the base of the crypt in the small intestine, where Paneth cells occupy the first three of four positions. In the mid region of the large bowel, the origin of the lineages can be traced to the very base of the crypt (Fig. 2). Until recently there were no effective stem cells specific markers that facilitated the study of these cells. However, the precise spatial organisation of the crypt meant that this was a unique cell biological model system where stem cells could be studied by virtue of their highly specific location in the tissue, as illustrated in Fig. 2. Amongst other things, this meant that the crypt model system could be used to provide in vivo analysis of the behaviour and response of stem cells to applied agents. Furthermore, comparisons could be made between the response of stem cells and stem cell-derived dividing transit cells to such agents. Typical effects on cell proliferation and apoptosis susceptibility could be studied. These parameters are the basis for commercial services provided by companies such as Epistem Ltd., i.e. preclinical in vivo assessment of the mode of action and toxicity/efficacy of novel drugs (Cycloquant and Apoquant).

An alternative hypothesis, which cannot be completely excluded, is that the ultimate steady state stem cells are located in the lower regions of the crypts scattered amongst the Paneth cells (Bjerknes and Cheng 1981, 2002; Cheng 1974). Certainly, undifferentiated primitive looking cells are located amongst the Paneth cells. However, the arguments against these being the ultimate stem cells come from the extrapolation of the

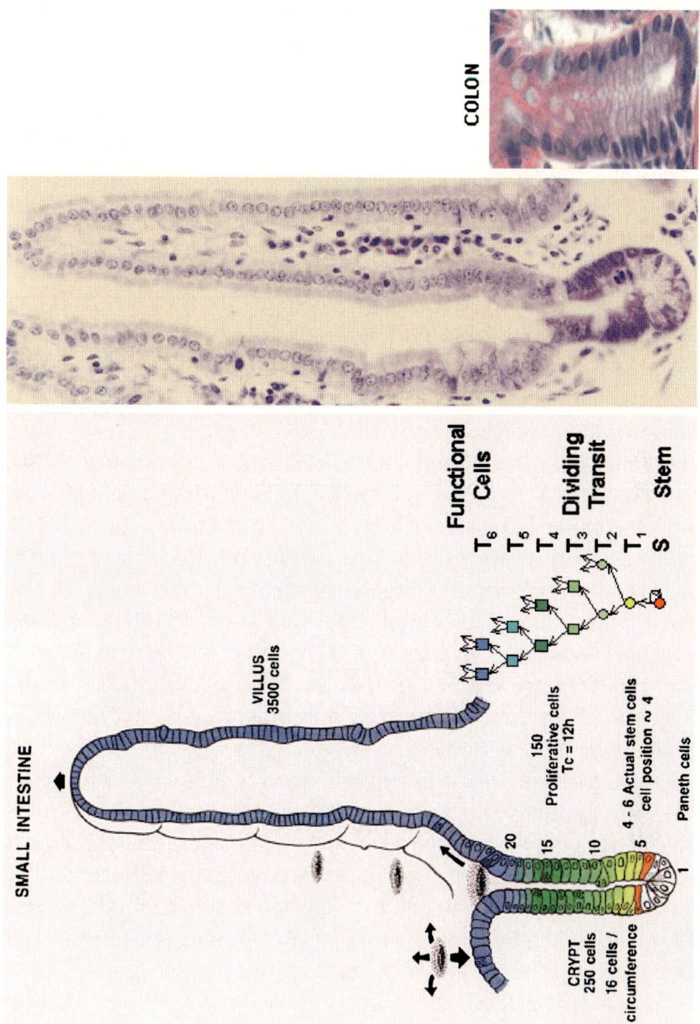

Fig. 2. Schematic representation of a section of small intestine (*left*) with a cell lineage on the *right*. The position of a cell in the lineage can be precisely related to the position in the section by the colour coding. The stem ells (*red*) can be studied by observing response at the fourth to fifth cell position. *Right panels*: Haematoxylin and eosin stained sections of small intestine and colon

velocity vs cell position data that have been published (Kaur and Potten 1986; Qiu et al. 1994), where the extrapolation clearly goes to the fourth or fifth position and also the fact that there are about five cells at the fourth or fifth position that have one or two specific characteristics that can be attributed to stem cells (see below).

The extensive background cell kinetic data and the more recent detailed cell positional data enabled one to determine that the dividing transit cells divide twice a day (i.e. they have a cell cycle time of 12 h). Moreover, one of the characteristic features of the cells at the stem cell position is that their cycle time is twice that of the transit cells (i.e. they have a cycle time of 24 h; Potten 1986). The cell cycles in the large bowel are generally about one and a half times slower in the mouse, and in humans, the cycle times are considerably slower in both areas of the intestine (Potten 1995).

If the stem cells in the small intestine of a mouse cycle once a day, this means that in the life span of a laboratory mouse, which might be 3 years, these cells divide a thousand times; a very large division potential. This has to be considered in the light of the average life expectancy of a mouse living in the wild of about 6 months. Surprisingly, the best estimate for humans would be that the small intestinal stem cells have a division potential of between 5,000 and 6,000 (Potten et al. 2003).

The rapid cell cycle of both stem and transit cells and the short life expectancy of the functional cells make this a tissue that is very susceptible to damage by genotoxic agents such as radiation and cytotoxic drugs, such as those used in chemotherapy. The effects of these agents is to kill, or reproductively sterilise, proliferating cells, which will result in a depletion of cellularity in the crypt and on the villus. If all the stem cells are reproductively sterilised, or killed, the cellular depletion in the crypts will be total and a micro-ulcer will develop. Such killing in groups of adjacent crypts will result in the appearance of macroscopically visible ulcers, such as observed post-irradiation and in cancer chemotherapy patients (gastrointestinal mucositis). If, however, one or more stem cells survive, a powerful stimulus for regeneration is triggered and a regenerated crypt will form which can be easily recognised in histological sections. Sterilized crypts disappear within 2–3 days of cytotoxic insult, and the regenerative foci or new crypts are easily recognisable on day 3 or day after exposure. This forms the basis of an in vivo stem cell survival and functional competence assay that is marketed by Epistem as Clonoquant. This is a direct in vivo measurement of stem cell survival, or its inverse, cell killing (Potten and Hendry 1985b) and revaluates the potential of a drug to cause gastrointestinal mucositis or, indeed, reduce the effects of mucositis induced by a known cytotoxic agent.

The fact that this tissue, by virtue of its rapid cell proliferation, is so sensitive to radiation has meant that extensive radiobiological studies have been undertaken and the Clonoquant assay (alternatively known as the crypt microcolony assay) has been fairly extensively studied. This microcolony assay is one of several in vivo stem cell survival assays that include the spleen colony assay for bone marrow stem cells and a range of other tissues collectively summarised in Potten and Hendry (1985a). A range of disparate observations (including stem cell survival, apoptosis at the stem cell location, DNA repair, mathematical modelling, studies into the initiation of regeneration and studies involving damaged response genes such as *p53* and survival genes such as *bcl-2*) have necessitated a modification of the cell lineage diagram shown in Fig. 2 to that shown in Fig. 3. The differences between these two models are subtle but have major implications in terms of stem cell behaviour. In the earlier model in Fig. 2, a differentiation event can be considered to occur at, or very close to, the cell division of the ultimate stem cells. This in fact implies that these cells divide asymmetrically under steady state conditions. The modification in Fig. 3 is that this differentiation event is postponed until the second or third generation down the lineage. The consequence of this is that all these disparate radiation observations can be accommodated. The model proposes a hierarchical stem cell compartment consisting of actual stem cells that function on a day-to-day basis and potential stem cells. If all actual stem cells are killed, the crypts do not disappear, because any one of the potential stem cells, which have not as yet made a commitment to differentiation, can reoccupy any vacant niche or location and function as future actual stem cells. A final consequence of this model is that these stem cells do not divide asymmetrically but go through a series of symmetric cell divisions. The feature that distinguishes actual from potential stem cells may solely be their location, or their association with a specific niche or environment.

The crypts are remarkably constant in size and one of the surprising and ill-understood processes is how the stem cell numbers are maintained

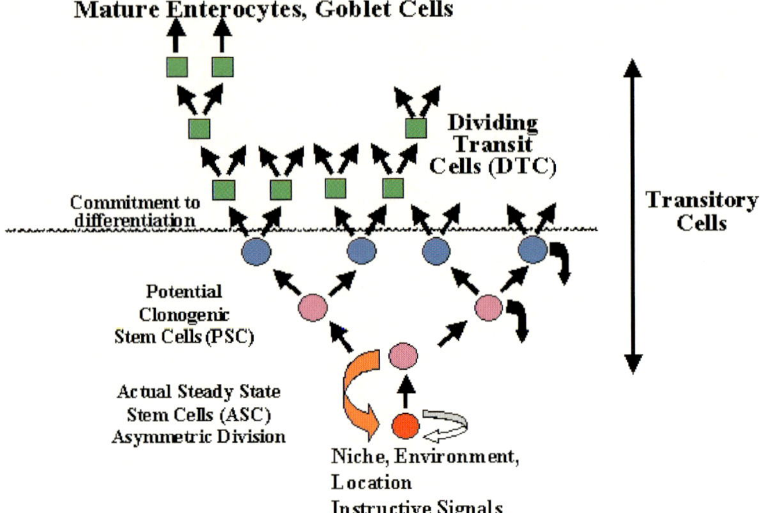

Fig. 3. Current working model of the small intestinal cell lineages showing a differentiation event distinguishing stem cells from stem cell progeny (dividing transit cells) at the third generation. All *circles* are stem cells but only the *red circles* function as stem cells under steady state conditions (i.e. are in the stem cell niche). Any circle (potential stem cell) can re-occupy a vacant niche

at a constant level. At cell position 4–5, there is an annulus of 16 cells that undulate as one goes round the circumference of the crypt between cell positions 2 and 7. This annulus is above the Paneth cell distribution, which is variable. Within this annulus it is believed there are five ultimate lineage ancestor stem cells. Each stem cell produces a lineage that results in 64–128 daughter cells. If one stem cell is lost, the crypt would shrink in size by between and 64 and 124. If an extra stem cell is produced by an occasional symmetric division, the crypt would increase in size, by the same amount. This does not appear to happen, which indicates that the number of stem cells is very tightly regulated (Marshman et al. 2002). The regulatory processes here are not understood but may involve a diffusible stem cell factor. The homeostatic regulatory process may account for the spontaneous apoptosis seen in normal mouse and human crypts, which may be removing occasional excess stem cells produced by occasional symmetric divisions or expansions of the niche.

The information that has been reviewed so far in this paper can be summarised as shown in Table 1. The data for mouse are fairly well established, whereas the information for humans is much more speculative.

One surprising feature associated with approximately five cells at the stem cell location in the small intestine is an exquisite intolerance of any genotoxic damage, manifested by the activation of an altruistic cell suicide or apoptosis, (Potten 1977; Hendry et al. 1982). This observation evolved out of attempts to measure the opposite of cell survival following radiation, i.e. cell death. With hindsight, it is clear that the crypt stem cell survival assay measures the survival of some relatively radioresistant potential clonogenic stem cells that have a very good capacity for repair of any sublethal radiation damage. It appears that these cells do not usually die by apoptosis but predominantly enter a premature differentiation (Paulus et al. 1992), i.e. persist in the tissue as cellular entities that lack a reproductive potential. The number of these potential clonogenic stem cells per crypt was somewhat dependent on the level of stress imposed, i.e. the levels of radiation damage, but ranges from approximately 6 to about 30 per crypt (Cai et al. 1997). It was on the basis of these calculations that the level of the differentiation event in the stem cell compartment in Fig. 3 was derived.

Table 1. Small intestinal crypts: a comparison between mouse and man

	Mouse	Man
No. of cells in a small intestinal crypt	250	~ 570
4–6 Lineages, Tc 12 h		
4–6 Stem cells, Tc 24 h		
Number of transit generations	~ 6	–
Villus transit time (days)	~ 3	–
No. Of divisions for each stem cell in a lifespan	1,000	$\sim 5,000 - 6,000$
Total number of crypts	7.5×10^5	5×10^7
Total number of stem cells	$\sim 3 - 4.5 \times 10^6$	$< 2 - 3 \times 10^8$
Total number of potential stem cells	$\sim 1 \times 10^8$	–
Total lifespan	2.25×10^{11}	2×10^{14}

Tc, duration of cell cycle

It turns out that apoptosis can be very rapidly initiated in a few cells at the stem cell position and that these exhibit the classical morphological changes ascribed to apoptosis as seen in electron microscopy. In good quality, fixed, sectioned and stained tissue, the apoptotic events can be clearly identified and counted (Potten 1977; Hendry et al. 1982). If this is done on a cell positional basis (see Apoquant) the highest frequency of this death occurs at cell positions 4–5. The cells die rapidly with peak levels being observed between 3 and 6 h of exposure. This death is totally P53-dependent (Merritt et al. 1994) and the dose response indicates an exquisite radiosensitivity. Indeed, it is the ultimate radiosensitivity theoretically possible if DNA is regarded as the target: one hit anywhere in the DNA triggers this apoptosis. Doses of the order of 1–5 cGy can raise the levels of apoptosis per crypt above the spontaneous P53-independent levels seen in both mouse and human crypts. There are a maximum of about five apoptotic susceptible cells at the fourth and fifth position, i.e. numbers and locations, suggesting that these are the ultimate actual stem cells. All of these cells are killed by a dose of 1 Gy, but the crypts are not reproductively sterilised: none of them disappear. They all regenerate, indicating that there are other regenerative stem cells that can repopulate the crypt. They do so rapidly and effectively, and within 2 days the actual radiation-susceptible stem cells are re-established (Ijiri and Potten 1984). Various observations suggest that the apoptosis-susceptible cells do not make use of DNA repair mechanisms but rather commit an altruistic cell suicide. The radiation studies including the apoptosis observations have been extensively reviewed; a recent example is Potten (2004).

Cancer of the large bowel is common in the Western world and its prognosis is poor. An interesting and somewhat ill-understood fact is that cancer of the small bowel is relatively rare. This is surprising bearing in mind the significantly larger mass of tissue in the small intestine and the fact that the level of proliferation is considerably higher. There are also considerably more potential carcinogen target cells (stem cells) in the small intestine; see Table 2 (see also Potten et al. 2003a). The carcinogen target cells, it could be argued, are most likely to be the small number of lineage ancestor cells upon which all cell replacement is ultimately dependent. All other proliferating cells have an extremely short life expectancy, measured in days (mouse) or weeks (humans), while the

Table 2. Small intestine in mouse and man compared with colon

3–4 times greater mass (length)
1.5 times faster proliferation
2–3 times more total stem cells
3–4 times more total stem cell divisions in a lifetime (4×10^9 in mouse and 1.4×10^{12} in man)

latent period between carcinogen exposure and cancer development can be months (mouse) or decades (humans).

Cancer development in the intestine is believed to be a multi-stage process with some genes, such as APC, playing a key gatekeeper role. It is possible that following three out of four genetic changes, with only one further change required for cancer development, all cells derived from the initial altered stem cell (i.e. its entire lineage) will carry the three premalignant alterations. The final change could then occur in any cell in the lineage, i.e. at any cell position. The development of a cancer from that cell, however, would not have occurred without the three proceeding changes occurring in a stem cell. There is also the issue of whether a cancer could develop in such a dynamic tissue where cell movement in the tightly packed epithelium is rapid and unidirectional. For a cancer to develop in such a cell it would be like a human on a packed moving staircase, stopping movement relative to the ground and at the same time proliferating! So the question of why we do not get more small intestinal cancers leads one to conclude that the stem cells in the small intestine must be extremely well protected against the accumulation of these genetic errors that lead to cancer.

DNA replication is one of the most genetically hazardous events that occur in the life of a stem cell and, as we have seen, these cells go through a very large number of rounds of DNA replication. A simple and complete protection against replication-induced errors would be to have evolved, in the stem cells, a mechanism for sorting the old template strands of DNA from the newly synthesised strands at division. The template strands would be retained in the daughter cell destined to become a stem cell. The newly synthesised strands with any replication errors would be acquired by the daughter cell destined to migrate, differentiate and ultimately be lost from the tissue. This hypothesis was proposed by

Cairns in 1975. In 1978 (Potten et al. 1978), data were produced that provided supporting evidence for both stem cells in the dorsal tongue epithelium (filiform papillae) and in the small intestinal crypts following radiation. In the latter case, the observations were based on the hypothesis that if stem cells were selectively segregating template strands from the newly synthesised strands, it ought to be possible to permanently label template strands in those circumstances where stem cells are making more stem cells and hence making new template strands. This occurs both during late development of the tissue (in the mouse small intestine about 6 weeks of age) and also following radiation injury where some stem cells have been killed forcing the other stem cells to repopulate the crypt with new stem cells. Experiments were performed in 1978 (Potten et al. 1978) using the radiation model and were extended in a more recent paper in 2002 (Potten et al. 2002) where both the postirradiation and the tissue developmental protocols were used. In both studies, permanently labelled cells located at the fourth or fifth position were observed. The labelling protocol involved tritiated thymidine given over a period of about 3 days. The numbers were lower than the predicted number of actual stem cells almost certainly as a consequence of technical difficulties. These permanently labelled cells could be seen to enter mitosis and could subsequently be labelled using another marker such as BrdU, when the first marker was tritiated thymidine. In this case, the BrdU would enter the newly synthesised strand in the permanently labelled cells, commonly called label-retaining cells (LRCs). In this way all the LRCs at the stem cell position in the crypt could become double-labelled with tritiated thymidine and BrdU. The experiments subsequently showed that these two labels segregated according to different kinetics. The tritiated thymidine persisted over a period of 10 days (equivalent to ten cell cycles) while the BrdU disappeared after 2 days (two cell cycles). As can be seen from Fig. 4, the Cairns hypothesis would predict that it takes two cell cycles to remove the BrdU from the LRCs.

These studies clearly show that the two labels segregate differentially so selective strand segregation must be operating in these rare but crucial stem cells. This has major implications in several areas of cancer biology and genetics and cell ageing (telomere shortening). An implied consequence of the Cairns hypothesis is that processes such as sister chromatid exchange must be prohibited in these cells since this would mix the two

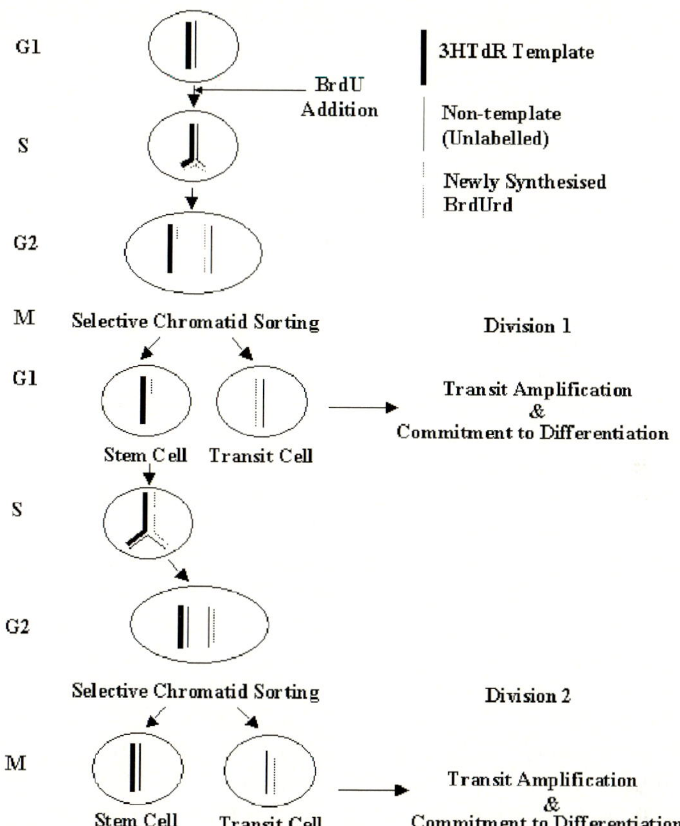

Fig. 4. Diagram showing the selective chromatid segregation ensuring retention of the template strand by stem cells. Retention of the old template strands (*thick solid line*) eliminates the risk of replication induced errors in the stem cells. Note that if a label is introduced into the newly synthesised strands (e.g. BrdU, *dotted line*) it takes two successive mitoses to remove that label from the stem cells

strands and defeat the purpose of the selective sorting. This might result in a compromised excision repair process since some enzymes would be common for both sister chromatid exchange and excision repair. As a consequence, it might be predicted that these cells would have a high insensitivity to genotoxic damage, i.e. a high radiosensitivity, which may provide the explanation for the apoptosis-susceptible cells discussed earlier. Therefore, these apoptotic susceptible cells may die as a consequence of the random genetic damage induced in the template strands.

Very similar observations have now been observed in mammary gland stem cells implanted in fat pads and prelabelled to generate LRCs. Using hormone stimulation of proliferation, newly synthesised strands can be labelled with BrdU and the segregation of the two labels in the doubly labelled cells studied. The number of LRCs remained constant but the BrdU label disappeared after 5 days from the doubly labelled cells (double labelling dropped from 83% of the LRCs to 15%) (Smith 2005).

The selective DNA segregation process also provides the most reasonable explanation for label-retaining cells seen in interfollicular epidermis, where the LRCs have various characteristics, including a location associated with stem cells.

There are also studies from cell lines where P53 expression is controlled by a zinc promoter, indicating a role for P53 in both asymmetric cell divisions and selective strand segregation (Merok et al. 2002). There are also other, as yet unpublished, stem cell systems that are exhibiting behaviour consistent with asymmetric strand segregation. Therefore, the Cairns segregation process may well be acting in the rare stem cells of several diverse tissues and thus may be a common feature of all ultimate tissue stem cells.

The mechanisms involved in the strand segregation remain unclear but may be related to new and old centriole formation, specific tubulins, centromeric attachments, etc. Some of the genes that may be associated with this process include P53, the aurora kinases, numb/notch, etc. It is unclear whether stem cells undergoing expansionary growth, i.e. symmetric divisions continue to sort old and new strands under situations where the new strands will ultimately be template strands in newly formed stem cells.

These observations that stem cells selectively sort old from new strands, and their altruistic apoptosis characteristics, provide a new explanation, at least in part, for the differential cancer incidence between the large and small bowel. It is not clear, because the experiments have not yet been conducted, whether the stem cells in the large bowel continue to selectively segregate old and new strands, but it is clear that in the large bowel, the altruistic apoptosis, which provides a secondary level of protection against template strand errors, is compromised by the expression of the anti-apoptotic cell survival gene bcl-2, (Merritt et al. 1995). Bcl-2 prevents both the naturally occurring apoptosis, which is believed to be part of the stem cell homeostatic mechanisms that ensure a stable stem cell population size per crypt, and also the genotoxic damage-induced apoptosis. Thus, in the large bowel, the size of the stem cell population, and hence ultimate carcinogen target population, may gradually drift upwards with the passage of time. These cells lack the ability to commit altruistic suicide if they sustain genotoxic damage and therefore represent an increase in the risk of cancer.

The high incidence of colorectal cancers has in the past been attributed in part to the differential bacterial flora in the small and large bowel and the potential chemical carcinogen bi-products from the higher bacterial load in the large intestine. One of the difficulties with this explanation is that if these chemical carcinogens are soluble enough to reach the susceptible target cells in the base of the crypt in the colon, they are equally likely to be absorbed into the blood stream and be distributed to other epithelial sites including the small intestine. The only chemical carcinogens that might have local specificity would be insoluble chemicals (less likely to be bacterial bi-products), which would find their way gradually to the base of the crypts, where they might be taken up by some of the carcinogen target stem cells. The soluble carcinogens are more likely to be absorbed at the surface of the colonic epithelium and distributed systemically.

Whether the stem cells in the large intestine operate the same selective DNA segregation process as observed in the small bowel remains to be investigated, as does the potential involvement of P53 in these segregation processes.

It is likely that some of the colonic bacterial bi-products are genotoxic and would trigger apoptosis in the stem cells. The concentration of these might be higher at the stem cell location in the colon than in the small intestine, whether considering the local bacterial load or the systemic delivery route. Evaluation might have selected to suppress apoptosis (via *bcl-2*) rather than have high levels and a constant regeneration (recall) of stem cells with any potential replication-induced errors. However, the latter would only tend to be errors associated with one, or very few, rounds of DNA replication. This was probably not important if the life span is short, but as medical advances increase the average life expectancy it may become more significant.

In conclusion, selective DNA segregation appears to be a common characteristic of stem cells in several tissues. It is a process by which stem cells can be permanently labelled, i.e. it is a way of marking stem cells and hence studying them and potentially isolating them. Other markers are currently under development and some of these were tentatively proposed in a recent review article (Potten 2004). One that has generated some interest has been Musashi 1, an RNA-binding protein associated with asymmetric division in neural stem cells (Potten et al. 2003b). The protein is expressed in a few cells in the stem cell location in the small intestine and increased expression is observed in conditions of stem cell expansion (i.e. in development and postradiation, and in early adenomas).

In general it can be said that stem cell biology is a rapidly expanding and topical field of scientific interest. Stem cells exist in all the adult tissues of the body. In at least five different tissues there is some evidence that the stem cells undergo selective strand segregation. A range of markers are being developed which will facilitate stem cell biology studies. Furthermore, our understanding of how to isolate and purify stem cells is evolving, and this is essential for the evolving field of tissue engineering. The major areas of research for the future include the identification of the key regulatory factors that determine stem cell proliferation rate, self-renewal probability, differentiation and cell death. Surprisingly, these key regulatory factors remain largely unknown for epithelial tissues and this is the basis of a major R&D programme within Epistem. Extensive stem cell regulation studies are required to develop the concept of tissue engineering and gene therapy where stem cell growth and expansion ex vivo are a prerequisite. There are major issues of debate at the moment concerning the plasticity of adult tissue stem cells and the ability to control the differentiation events that arise from pluri- or toti-potent stem cells, i.e. adult or embryonic stem cells. There is a great need for further work in the area of the stem cell involvement, in various diseases, most notably cancer and also the role of stem cells in tissue ageing (tissues may age because of a depletion in stem cell numbers or because of malfunctions in the regulation and damage response mechanisms).

References

Bjerknes M, Cheng H (1981) The stem-cell zone of the small intestinal epithelium. III. Evidence from columnar, enteroendocrine, and mucous cells in the adult mouse. Am J Anat 160:77–91

Bjerknes M, Cheng H (2002) Multipotential stem cells in adult mouse gastric epithelium. Am J Physiol Gastrointest Liver Physiol 283:767–777

Cai W, Roberts SA, Potten CS (1997) The number of clonogenic cells in three regions of murine large intestinal crypts. Int J Rad Biol 71:573–579

Cairns J (1975) Mutation selection and the natural history of cancer. Nature 255:197–200

Cheng H (1974) Origin, differentiation and renewal of the four main epithelial cell types in the mouse small intestine. V. Unitarian theory of the origin of the four epithelial cell types. Am J Anat 141:537–562

Hendry JH, Potten CS, Chadwick C, Bianchi M (1982) Cell death (apoptosis) in the mouse small intestine after low doses: effects of dose-rate, 14.7 MeV neutrons, and 600 MeV (maximum energy) neutrons. Int J Radiat Biol Relat Stud Phys Chem Med 42:611–620

Ijiri K, Potten CS (1984) The reestablishment of hypersensitive cells in the crypts of irradiated mouse intestine. Int J Radiat Biol 46:609–623

Kaur P, Potten CS (1986) Circadian variation in migration velocity in small intestinal epithelium. Cell Tissue Kinet 19:591–599

Marshman E, Booth C, Potten CS (2002) The intestinal epithelial stem cell. (Our favourite cell). Bioessays 24:91–98

Meineke FA, Potten CS, Loeffler M (2001) Cell migration and organization in the intestinal crypt using a lattice-free model. Cell Prolif 34:253–266

Merok JR, Lansita JA, Tunstead JR, Sherley JL (2002) Cosegregation of chromosomes containing immortal DNA strands in cells that cycle with asymmetric stem cell kinetics. Cancer Res 162:6791–6795

Merritt AJ, Potten CS, Hickman JA, Kemp C, Balmain A, Hall P, Lane D (1994) The role of p53 in spontaneous and radiation-induced intestinal cell apoptosis in normal and p53 deficient mice. Cancer Res 54:614–617

Merritt AJ, Potten CS, Watson AJ, Loh DY, Nakayama K, Hickman JA (1995) Differential expression of bcl-2 in intestinal epithelia. Correlation with attenuation of apoptosis in colonic crypts and the incidence of colonic neoplasia. J Cell Sci 108:2261–2271

Paulus U, Potten CS, Loeffler M (1992) A model of the control of cellular regeneration in the intestinal crypt after perturbation based solely on local stem cell regulation. Cell Prolif 25:559–578

Potten CS (1977) Extreme sensitivity of some intestinal crypt cells to X and gamma irradiation. Nature 269:518–521

Potten CS (1986) Cell cycles in cell hierarchies. Int J Radiat Biol 49:257–278

Potten CS (1995) Structure, function and proliferative organisation of mammalian gut. In: Potten CS, Hendry JH (eds) Radiation and gut. Elsevier, Amsterdam, pp 1–31

Potten CS (1998) Stem cells in gastrointestinal epithelium: numbers, characteristics and death. Phil Trans Roy Soc Lond B 353:821–830

Potten CS (2004) Weiss lecture: radiation, the ideal cytotoxic agent for studying the cell biology of tissues such as the small intestine. Radiat Res 161:123–136

Potten CS, Loeffler M (1990) Stem cells: attributes, cycles, spirals, pitfalls and uncertainties. Lessons for and from the crypt. Development 110:1001–1020

Potten CS, Hendry JH (1985a) Cell clones: manual of mammalian cell techniques. In: Potten CS, Hendry JH (eds) Churchill Livingstone, Edinburgh

Potten CS, Hendry JH (1985b) The microcolony assay in mouse small intestine. In: Potten CS, Hendry JH (eds) Cell clones: manual of mammalian cell techniques. Churchill Livingstone, Edinburgh, pp 50–60

Potten CS, Hume WJ, Reid P, Cairns J (1978) The segregation of DNA in epithelial stem cells. Cell: 15:899–906

Potten CS, Booth C, Pritchard DM (1997) The intestinal epithelial stem cell: the mucosal governor. Int J Exp Path 78:219–243

Potten CS, Owen G, Booth D (2002) Intestinal stem cells protect their genome by selective segregation of template DNA strands. J Cell Sci 115:2381–2388

Potten CS, Booth C, Hargreaves D (2003a) The small intestine as a model for evaluating adult tissue stem cell drug targets. Cell Proliferat 36:115–129

Potten CS, Booth C, Tudor GL, Booth D, Brady G, Hurley P, Ashton G, Clarke R, Sakakibara S, Okano H (2003b) Identification of a putative intestinal stem cell and early lineage marker: musashi-1. Differentiation 71:28–41

Smith GH (2005) Label-retaining epithelial cells in mouse mammary gland divide asymmetrically and retain their template DNA strands. Development 132:681–687

Qiu JM, Roberts SA, Potten CS (1994) Cell migration in the small and large bowel shows a strong circadian rhythm. Epithelial Cell Biol 3:137–148

Winton DJ, Ponder BA (1990) Stem-cell organization in mouse small intestine. Proc Biol Sci 241:13–18

Winton DJ, Blount MA, Ponder BA (1998) A clonal marker induced by mutation in mouse intestinal epithelium. Nature 333:463–466

Wong MH, Saam JR, Stappenbeck TS, Rexer CH, Gordon JI (2000) Genetic mosaic analysis based on Cre recombinase and navigated laser capture microdissection. Proc Natl Acad Sci U S A 97:12601–12606

8 Tracking Stem Cells In Vivo

R. Yoneyama, E.R. Chemaly, R.J. Hajjar

8.1   Introduction 100
8.2   Criteria for Ideal Stem Cell In Vivo Imaging Technology
        and Potential Obstacles 100
8.3   In Vivo Stem Cell Tracking Methods 101
8.3.1   Optical Imaging: Bioluminescence and Fluorescence 101
8.3.2   Ultrasound 102
8.3.3   Single Photon Emission Computed Tomography 102
8.3.4   Positron Emission Tomography 103
8.3.5   Magnetic Resonance Imaging 104
8.3.6   X-Ray-Based Methods: Plain Films and Computed Tomography . 106
8.3.7   Multimodality Imaging 106
8.3.8   Histologic Validation and Laser Capture Microdissection
        of Stem Cells 106
8.4   Conclusion and Future Directions 106
References 107

Abstract. Stem cells have been targeted to many organ systems specifically to replace scarred organs and to rejuvenate diseased organs. Even though our understanding of the versatility of stem cells is slowly unraveling, tracking these cells as they enter the body has become a very important field of study. In this chapter, we review various modalities for imaging stem cells and assess the advantages and shortcomings of each technique.

8.1   Introduction

Stem cells are used to treat various types of cardiovascular disease. Specifically, they have been used to regenerate myocardium lost because of damage from myocardial infarction or to induce vasculogenesis in ischemic areas. Successful myocardial regeneration from stem cell differentiation has reportedly been achieved in animal studies (Orlic et al. 2001). Also, human clinical trials have shown improved myocardial contractile function, remodeling, and perfusion after endothelial progenitor cell and bone marrow-derived cell injection in patients with acute myocardial infarctions (Britten et al. 2003; Strauer et al. 2002). Improvement of contractile function and myocardial perfusion has also been achieved in end-stage refractory ischemic heart failure patients by injecting autologous bone marrow cells into ischemic areas with decreased contractility (Perin et al. 2003).

However, the fate, function, and movement of these cells once they are injected have not been well defined; therefore their specific contribution to regeneration, perfusion, and contractile function improvement is still poorly understood. For this reason, it has become imperative to track stem cells in an effective way to better assess their movement, proliferation, and contribution to improvement in contractile function following cell transfer in ischemic hearts (Orlic et al. 2001; Saito et al. 2003). In addition, the advent of molecular imaging technology applicable to stem cell tracking has expanded the possibilities for imaging stem cells (Wickline et al. 2003).

8.2   Criteria for Ideal Stem Cell In Vivo Imaging Technology and Potential Obstacles

Specific characteristics of an ideal imaging platform for stem cells targeting the cardiovascular system include being able to detect a small number of cells, to achieve single cell sensitivity, and to enable cell quantification. In addition, the technology must be safe and the contrast agent must be biocompatible and nontoxic, and ideally, would also allow long-term tracking without additional contrast administration.

Furthermore, the contrast agent, incorporated into the stem cells, must not dilute with cell division or transfer to non-stem cells, especially after cell death. These dilution/transfer problems are common to a wide range of agents presented in this review, such as near-infrared (NIR) fluorophores, contrast agents used for ultrasound, and iron oxide nanoparticles used in magnetic resonance imaging (MRI). In addition, diffusion of these agents can result in interstitial deposition of the contrast agent, which then may be misrepresented as proof of target cell bioactivity (Hill et al. 2003). Some of these problems can be overcome by genetic modification of the stem cell to generate the tracking signal. On one hand, this approach generates additional cost and has the potential for genetic instability with resulting safety issues (Hacein-Bey-Abina et al. 2003). On the other, in an experimental study involving rat bone marrow stromal cells injected after myocardial infarction, cardiomyogenic differentiation and improvement in myocardial contractility were observed, despite transfection of the cells with LacZ reporter gene for labeling (Saito et al. 2003).

This review presents the current methods and their advantages and limits. No current imaging technology fulfills all the above requirements (reviewed in Frangioni et al. 2004).

8.3   In Vivo Stem Cell Tracking Methods

8.3.1   Optical Imaging: Bioluminescence and Fluorescence

Bioluminescence detection uses the visible light generated by the enzyme luciferase (Wu et al. 2003). This technique has many limitations that preclude human use, including the absorption and scatter of visible light in tissue (limiting its use to small animals), the need for non-human gene expression (luciferase), and the need for injection of potentially immunogenic non-human substrates.

Fluorescence imaging uses fluorophores as exogenous contrast agents. Due to absorption and scatter of photons at visible wavelengths, near-infrared (NIR) fluorophores would seem to have the highest potential. In the NIR wavelength window (700–900 nm) and as opposed to visible and infrared wavelengths, absorbance spectra for all biomolecules reach minima (Chance 1998). In an experimental protocol using a NIR-fluorescent bisphosphonate for small animal imaging of bone hydroxyapatite, Zaheer et al. have shown high sensitivity and spatial resolution along with improved visualization of deep structures and visualization of NIR fluorescence by microscopy, allowing histological ex vivo validation of in vivo imaging (Zaheer et al. 2001).

8.3.2 Ultrasound

Tracking of stem cells for cardiovascular clinical trials by echocardiography is an attractive option, especially since echocardiography with contrast can achieve single cell detection (Klibanov et al. 2004). A variety of contrast agents have been developed for ultrasound (reviewed in Morawski et al. 2005), including echogenic nanoparticles, liposomes, and microbubbles. Some ultrasound contrast agents have already been used to target cell membrane receptors and extracellular molecules (Morawski et al. 2005). The first reported molecular imaging agents for ultrasound were targeted perfluorocarbon nanoparticles targeting fibrin thrombi and, in addition, in vivo vascular epitopes in smooth muscle cells (Wickline et al. 2003). Limitations of echocardiography and echocardiographic contrast agents include accuracy of cell quantification, spatial resolution, large contrast-agent particle size (limiting contrast, in most cases, to the vasculature), and lack of "robust" techniques for intracellular accumulation of the agent.

8.3.3 Single Photon Emission Computed Tomography

Single photon emission computed tomography (SPECT) detects high-energy gamma rays emitted by radioactive atoms (e.g., ^{99m}Tc) by rotating a collimated gamma camera around the subject and reconstructing tomographic images images. Different strategies for stem cell detection have been reported, as follows:

1. Direct loading of the stem cells with a radiometal through incubation in the cell culture medium permits tracking of stem cells and their body distribution after injection into rats with myocardial infarction (Barbash et al. 2003). Limitations include the trade-off between half-life and long-term exposure, and possible transfer of radioactivity to non-stem cells.
2. Enzymatic conversion and retention of a radioactive substrate, used in SPECT and positron emission tomography (PET), involves intro-

duction of the enzyme through a transgene, which allows reliable indefinite tracking after integration but also requires genetic manipulation and substrate injection at each imaging session. This approach (reviewed in MacLaren et al. 2000) was first developed for the purpose of transgene imaging, initially using visible light reporters and subsequently PET probes. In vivo imaging in deep tissue is possible using high-energy gamma rays with possible quantification and a high degree of sensitivity compared to optical imaging and MRI (MacLaren et al. 2000). An important consideration is the gene chosen for contrast generation: endogenous genes generate background signal and exogenous gene products are potentially immunogenic. Herpes simplex virus 1 thymidine kinase is an enzymatic PET reporter gene with therapeutic potential (MacLaren et al. 2000).

3. Stable expression (in the stem cells) of a receptor (e.g., D2R) and its binding to a radiolabeled ligand (MacLaren et al. 2000; Simonova et al. 2003) are also possible. An alternative is the use of an endogenous receptor and a radiolabeled ligand. For example, a recent study used radiolabeled transferrin for the tracking of human mesenchymal stem cells injected into rabbit spinal chord natively expressing the transferrin receptor at higher levels than the surrounding tissue, which resulted in specific radioscintigraphic tracking (Ding et al. 2004).

4. Radiolabeled oligonucleotide can bind a mRNA transcript (MacLaren et al. 2000).

5. Radiolabeled antibodies/antibody-like structures can target cell-surface proteins (MacLaren et al. 2000).

8.3.4 Positron Emission Tomography

PET uses coincident detection of two antiparallel gamma rays emitted after positron annihilation and is overall more sensitive than SPECT with more accurate quantification of cell number. The same detection strategies described above for SPECT and reviewed in other articles (Frangioni et al. 2004; MacLaren et al. 2000) can be used with PET.

Problems with PET and SPECT include nonspecific uptake of the tracer by normal tissue, non-negligible tissue photon attenuation, and the need for an adequately high cellular concentration of tracer to meet detection thresholds. Hybrid systems combining CT and PET/SPECT have been used for attenuation correction and combine anatomic and physiologic imaging (Frangioni et al. 2004).

8.3.5 Magnetic Resonance Imaging

MRI is the most used utilized modality for in vivo stem cell tracking mode due to its safety and three-dimensional capabilities (Bulte et al. 2001). Contrast generation is via either T1 or T2/T2* mechanisms.

T1 agents utilize gadolinium to shorten the T1 relaxivity of protons from associated water molecules and increase the signal on T1-weighted images; imaging of stem cells is possible through gadolinium attached to scaffolds such as dendrimers or dextrans and loaded in cells through endocytosis (Modo et al. 2002).

T2/T2* contrast is typically generated using superparamagnetic iron oxide nanoparticles loaded in endosomes via endocytosis. The endocytosed particles make susceptibility artifacts around themselves, which magnifies their T2/T2* shortening effect and therefore permits small numbers of stem cells to be tracked (reviewed in Frangioni et al. 2004).

A notable strategy used to improve cell incorporation of magnetic particles is the HIV-Tat peptide. In a published study, use of Tat-loading permitted MRI tracking of hematopoietic progenitors at a single cell level, and cells were harvested through magnetic separation (Lewin et al. 2000).

Several FDA-approved T1 and T2/T2* contrast agents were used successfully to target human hematopoietic progenitor cells, with a higher sensitivity for T2 (Daldrup-Link et al. 2003). The feasibility of MRI tracking after injection of Feridex-labeled mesenchymal stem cells in a pig myocardial infarction model has been demonstrated (Kraitchman et al. 2003). A study where similar cells were injected in normal pig hearts showed the feasibility of long-term tracking (up to 7 weeks) without impairment of myocardial function (Yoneyama et al. 2004) is shown in Fig. 1. Another study of serial MRI of mesenchymal stem cells after intracardiac injection in a swine myocardial infarction model with the cells stained with iron fluorescent particles showed the feasibility of serial T2/T2*-weighted MR imaging and fluoroscent imaging of the tissue while preserving in vitro cell viability and adipogenic and osteogenic

Feridex Labeled Stem Cells

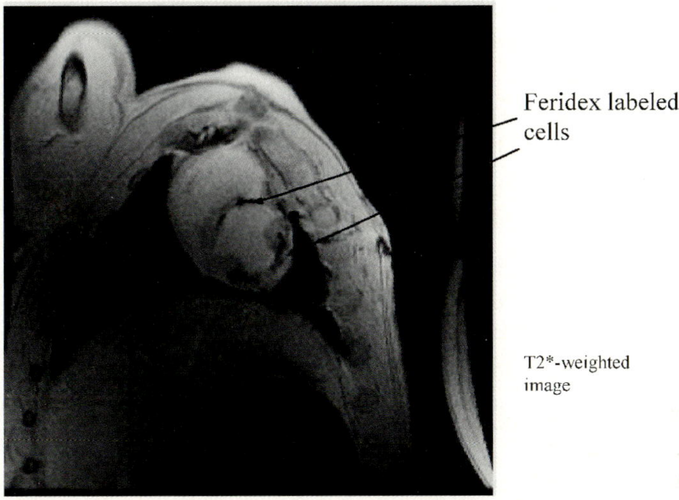

Fig. 1. Bone marrow-derived mesenchymal cells (*MSC*) were injected into the left ventricle of a normal ventricle. Seven weeks later, T2*-weighted image showed the presence of the MSCs labeled with Feridex

differentiation (Hill et al. 2003). The ultimate sensitivity of MRI with respect to stem cell number, using any of the above strategies, remains to be determined.

Incorporation of MRI contrast agents did not affect the viability of human hematopoietic progenitors (Daldrup-Link et al. 2003). Also, another study on magnetodendrimers showed that their incorporation did not affect the differentiation of neural stem cells (Bulte et al. 2001). However, a recent report has raised concerns in that regard by showing the inability of mesenchymal stem cells loaded with Feridex to undergo chondrogenic differentiation (Kostura et al. 2004). The latter issue, however, remains controversial in the presence of another study with contradictory results but different assessment methods (Arbab et al. 2004; Bulte et al. 2004).

MRI was also evaluated as a tool for guiding gene and cell transfer in a rat model of myocardial infarction (Barbash et al. 2004).

8.3.6 X-Ray-Based Methods: Plain Films and Computed Tomography

Since these techniques require high and thus difficult to achieve concentrations of contrast material (e.g., iodine, gadolinium) in imaged cells, they have little role in stem cell tracking at the present time (Frangioni et al. 2004).

8.3.7 Multimodality Imaging

Multimodality contrast agents have been used to overcome the limitations of individual strategies, combining optical, MRI, and ultrasound contrast (Frangioni et al. 2004).

8.3.8 Histologic Validation and Laser Capture Microdissection of Stem Cells

A number of studies have reported microscopic visualization of the imaging contrast agent loaded in stem cells (Hill et al. 2003; Kraitchman et al. 2003; Yoneyama et al. 2004). Laser capture microdissection (LCM) was able to retrieve stem cells in the mouse intestinal epithelium and their mRNA could be analyzed by reverse-transcription PCR (Wong et al. 2000), expanding the possibilities of stem cell tracking.

8.4 Conclusion and Future Directions

Given the various features and limitations of stem cell tracking techniques, the ideal (i.e., safe, available, sensitive, accurate, and convenient) technique is not yet available. Currently, the individual imaging modalities can provide specific answers in terms of stem cell presence, movement, and viability, while robust quantification and tracking are still lacking. Therefore, multimodality imaging and other novel approaches have the potential to provide better stem cell monitoring (reviewed in Frangioni et al. 2004).

References

Arbab AS, Yocum GT, Kalish H et al. (2004)Efficient magnetic cell labeling with protamine sulfate complexed to ferumoxides for cellular MRI. Blood 104:1217–1223

Barbash IM, Chouraqui P, Baron J et al. (2003) Systemic delivery of bone marrow-derived mesenchymal stem cells to the infarcted myocardium: feasibility, cell migration, and body distribution. Circulation 108:863–868

Barbash IM, Leor J, Feinberg MS et al. (2004) Interventional magnetic resonance imaging for guiding gene and cell transfer in the heart. Heart 90:87–91

Britten MB, Abolmaali ND, Assmus B et al. (2003) Infarct remodeling after intracoronary progenitor cell treatment in patients with acute myocardial infarction (TOPCARE-AMI): mechanistic insights from serial contrast-enhanced magnetic resonance imaging. Circulation 108:2212–2218

Bulte JW, Douglas T, Witwer B et al. (2001) Magnetodendrimers allow endosomal magnetic labeling and in vivo tracking of stem cells. Nat Biotechnol 19:1141–1147

Bulte JW, Kraitchman DL, Mackay AM, Pittenger MF (2004) Chondrogenic differentiation of mesenchymal stem cells is inhibited after magnetic labeling with ferumoxides. Blood 104:3410–3412; author reply 3412–3413

Chance B (1998) Near-infrared images using continuous, phase-modulated, and pulsed light with quantitation of blood and blood oxygenation. Ann N Y Acad Sci 838:29–45

Daldrup-Link HE, Rudelius M, Oostendorp RA et al. (2003) Targeting of hematopoietic progenitor cells with MR contrast agents. Radiology 228:760–767

Ding W, Bai J, Zhang J et al. (2004) In vivo tracking of implanted stem cells using radio-labeled transferrin scintigraphy. Nucl Med Biol 31:719–725

Frangioni JV, Hajjar RJ (2004) In vivo tracking of stem cells for clinical trials in cardiovascular disease. Circulation 110:3378–3383

Gimbel JR, Kanal E (2004) Can patients with implantable pacemakers safely undergo magnetic resonance imaging? J Am Coll Cardiol 43:1325–1327

Hacein-Bey-Abina S, Von Kalle C, Schmidt M et al. (2003) LMO2-associated clonal T cell proliferation in two patients after gene therapy for SCID-X1. Science 302:415–419

Hill JM, Dick AJ, Raman VK et al. (2003) Serial cardiac magnetic resonance imaging of injected mesenchymal stem cells. Circulation 108:1009–1014

Klibanov AL, Rasche PT, Hughes MS et al. (2004) Detection of individual microbubbles of ultrasound contrast agents: imaging of free-floating and targeted bubbles. Invest Radiol 39:187–195

Kostura L, Kraitchman DL, Mackay AM, Pittenger MF, Bulte JW (2004) Feridex labeling of mesenchymal stem cells inhibits chondrogenesis but not adipogenesis or osteogenesis. NMR Biomed 17:513–517

Kraitchman DL, Heldman AW, Atalar E et al. (2003) In vivo magnetic resonance imaging of mesenchymal stem cells in myocardial infarction. Circulation 107:2290–2293

Lewin M, Carlesso N, Tung CH et al. (2000) Tat Peptide-derivatized magnetic nanoparticles allow in vivo tracking and recovery of progenitor cells. Nat Biotechnol 18:410–414

MacLaren DC, Toyokuni T, Cherry SR et al. (2000) PET imaging of transgene expression. Biol Psychiatry 48:337–348

Martin ET, Coman JA, Shellock FG, Pulling CC, Fair R, Jenkins K (2004) magnetic resonance imaging and cardiac pacemaker safety at 1.5-Tesla. J Am Coll Cardiol 43:1315–1324

Modo M, Cash D, Mellodew K et al. (2002) Tracking transplanted stem cell migration using bifunctional, contrast agent-enhanced, magnetic resonance imaging. Neuroimage 17:803–811

Morawski AM, Lanza GA, Wickline SA (2005) Targeted contrast agents for magnetic resonance imaging and ultrasound. Curr Opin Biotechnol 16:89–92

Orlic D, Kajstura J, Chimenti S et al. (2001) Bone marrow cells regenerate infarcted myocardium. Nature 410:701–705

Perin EC, Dohmann HF, Borojevic R et al. (2003) Transendocardial, autologous bone marrow cell transplantation for severe, chronic ischemic heart failure. Circulation 107:2294–2302

Roguin A, Donahue JK, Bomma CS, Bluemke DA, Halperin HR (2005) Cardiac magnetic resonance imaging in a patient with implantable cardioverter-defibrillator. Pacing Clin Electrophysiol 28:336–338

Saito T, Kuang JQ, Lin CC, Chiu RC (2003) Transcoronary implantation of bone marrow stromal cells ameliorates cardiac function after myocardial infarction. J Thorac Cardiovasc Surg 126:114–123

Simonova M, Shtanko O, Sergeyev N, Weissleder R, Bogdanov A Jr (2003) Engineering of technetium-99m-binding artificial receptors for imaging gene expression. J Gene Med 5:1056–1066

Strauer BE, Brehm M, Zeus T et al. (2002) Repair of infarcted myocardium by autologous intracoronary mononuclear bone marrow cell transplantation in humans. Circulation 106:1913–1918

Wickline SA, Lanza GM (2003) Nanotechnology for molecular imaging and targeted therapy. Circulation 107:1092–1095

Wong MH, Saam JR, Stappenbeck TS, Rexer CH, Gordon JI (2000) Genetic mosaic analysis based on Cre recombinase and navigated laser capture microdissection. Proc Natl Acad Sci U S A 97:12601–12606

Wu JC, Chen IY, Sundaresan G et al. (2003) Molecular imaging of cardiac cell transplantation in living animals using optical bioluminescence and positron emission tomography. Circulation 108:1302–1305

Yoneyama R, Pomerantseva I, Kawase Y et al. (2004) Magnetic resonance imaging of ferumoxide labeled mesenchymal stem cells in myocardium injected by catheter-based technique via coronary vein. The Tony and Shelly Malkin Stem Cell Symposium November 8, 2004. Boston, MA, USA

Zaheer A, Lenkinski RE, Mahmood A, Jones AG, Cantley LC, Frangioni JV (2001) In vivo near-infrared fluorescence imaging of osteoblastic activity. Nat Biotechnol 19:1148–1154

9 Establishment of Nuclear Transfer Embryonic Stem Cell Lines from Adult Somatic Cells by Nuclear Transfer and Its Application

T. Wakayama

9.1 Introduction 112

9.2 Establishment of Nuclear Transfer Embryonic Stem Cell Lines
from Different Mice 113

9.3 Application of Nuclear Transfer Embryonic Stem Cell Techniques 115

9.4 Preservation of Genes from Infertile Mice
Without the Use of Germ Cells 116

9.5 Nuclear Transfer Embryonic Stem Cell Derivation
from Lethal Cloned Embryos 118

9.6 Perspectives 119

References 120

Abstract. Nuclear transfer can be used to generate embryonic stem cell (ntESC) lines from a patient's own somatic cells. We have shown that ntESCs can be generated relatively easily from a variety of mouse genotypes and cell types of both sexes, even though it may be more difficult to generate clones directly. Several reports have already demonstrated that ntESCs can be used in regenerative medicine in order to rescue immunodeficient or infertile phenotypes. However, it is unclear whether ntES cells are identical to fertilized embryonic stem cells (ESCs). This review seeks to describe the phenotype and possible abnormalities of ntESC lines.

9.1 Introduction

There has been a great deal of interest in the possible application of human embryonic stem cells (ESCs) in regenerative medicine, as ESCs are envisioned to be potential sources for cell replacement therapies. However, as with any allogeneic material, ESCs derived from fertilized blastocysts and the progeny of such cells inevitably face the risk of immunorejection on transplantation. It has been proposed that ESCs derived from embryos cloned from the host patient's own cell nuclei represent a potential solution to the problem of rejection, as any replacement cells would be genetically identical to the host's own somatic cell nuclei (Gurdon and Colman 1999; Mombaerts 2003; Wakayama 2004). These ES-like cell lines generated from somatic cells via nuclear transfer embryonic stem cells (ntESCs) were first reported in the fetus cow (Cibelli et al. 1998). Reports in the adult mouse (Munsie et al. 2000; Kawase et al. 2000; Wakayama et al. 2001) and adult cow (Wang et al. 2005) followed. These ntESC lines are believed to possess the same capacities for unlimited differentiation and self-renewal as conventional ESC lines derived from normal embryos produced by fertilization. We have previously shown that ntESC lines are capable of differentiating into all three germ layers in vitro or into spermatozoa and oocytes in chimeric mice (Wakayama et al. 2001). This was the first demonstration that ntESCs have the same developmental potential as ESCs from fertilized blastocysts and cloned mice can be obtained from these ntESC lines using a second nuclear transfer (Wakayama et al. 2001). These techniques have now been applied to not only preliminary medical research (Rideout et al. 2002), but also to basic biological research. For example, irreversible changes were demonstrated in the DNA of adult lymphocytes (Hochedlinger and Jaenisch 2002) but not of olfactory neurons (Eggan et al. 2004; Li et al. 2004). Cancer cell types have also been characterized using this technique (Blelloch et al. 2004; Hochedlinger et al. 2004). We also demonstrated that this technique is applicable to the preservation of genetic resources of mouse strains instead of the embryo, oocyte, or spermatozoan. At present, this technique is the only available method for the preservation of valuable genetic resources from mutant mice without the use of germ cells (Wakayama et al. 2005a).

On the other hand, the rate of successful production of cloned animals is still very low. Cloned embryos could develop to full term, although most animals show abnormalities of gene expression in the embryo (Boiani et al. 2002; Bortvin et al. 2003) and placenta (Wakayama and Yanagimachi 1999a; Tanaka et al. 2001), obesity (Tamashiro et al. 2000; 2002), and early embryonic death (Ogonuki et al. 2002). These are thought to be due to failures in genomic reprogramming, but the mechanism is poorly understood. ntESC lines showed an almost identical phenotype to normal ESCs, i.e., positive Oct3/4 and SSEA-1 expression (unpublished observation). However, it is still unclear if ntESCs possess the same abnormalities as cloned animals. If ntESCs are not normal, it might be dangerous to use regenerative medicine. This review seeks to describe the phenotype and possible abnormalities of ntESC lines.

9.2 Establishment of Nuclear Transfer Embryonic Stem Cell Lines from Different Mice

We conducted investigations using mouse somatic cell nuclei. In the mouse, the genotype of the animal strain or the sex of the donor nucleus often affects the successful full-term development of cloned offspring (Wakayama and Yanagimachi 2001; Inoue et al. 2003). For example, hybrid genotypes are more tolerant of cloning than inbred genotypes, such as C57BL/6 and C3H/He, which are commonly used in mouse genetic studies but have never been cloned successfully.

In order to determine the efficiency of ntESC line establishment, we have tested several different male and female mouse genotypes as sources of nuclei (Table 1). Nuclei of ntESCs were injected into recipient enucleated oocytes from adult hybrid mice, and ntESC lines were successfully established from all mouse genotypes of both males and females. However, when inbred and F_1 genotypes were compared, the rate of development to the blastocyst stage and the frequency of ntESC derivation were significantly better when F_1 cumulus cells were used as the donor nucleus. However, this difference was only seen when the data were compared from reconstructed oocytes. No strain- or gender-dependent differences were detected in tail-tip cells (Wakayama et al. 2005b).

Table 1. Establishment of ntESC lines from different mouse strains and organs

Mouse strain	Type of donor cell	Sex	No. of established ntES cell lines	% From morula/ blastocyst	Confirmation of germ line transmission
B6D2F1	Tail	Male	11	26.8	Yes
	Tail	Female	4	20.0	Yes
	Cumulus	Female	8	16.3	Yes
B6C3F1	Tail	Male	5	21.7	Yes
	Tail	Female	1	33.3	Yes
	Cumulus	Female	27	19.3	ND
BD129F1	Tail	Male	31	58.5	ND
	Cumulus	Female	12	75.0	ND
C3H/He	Tail	Male	1	33.3	Yes
	Tail	Female	2	16.7	Yes
	Cumulus	Female	1	5.9	Yes
C57BL/6	Tail	Male	2	14.3	Yes
	Cumulus	Female	5	9.8	Yes
DBA/2	Tail	Male	2	18.2	Yes
	Cumulus	Female	2	10.0	Yes
129/Sv	Tail	Male	1	2.7	Yes
	Cumulus	Female	1	7.7	Yes
C57BL/6$^{nu/nu}$	Tail	Male	4	4.5	ND
	Tail	Female	5	6.7	Yes
FVB	Tail	Male	3	23.1	ND
ICR	Tail	Male	1	9.1	ND
	Tail	Female	3	23.1	Yes
Tg-BDF1	Tail	Male	3	20.0	Yes
	Cumulus	Female	3	12.0	Yes
129B6F1	Sertoli	Male	4	26.7	Yes
Total			142	23.3	

In the first reports of cloning in mice (Wakayama et al. 1998) and cows (Kato et al. 1998), female cumulus cells were used as the donor nucleus. Thus, cumulus cells prevailed in nuclear transfer experiments, but cumulus cells are available only in the case of a healthy and fertile female. On the other hand, fibroblastic cells are abundantly and ubiquitously obtained from both females and males, and even unhealthy animals. We also compared the efficiency of ntESC establishment between fibroblast and cumulus cells and male and female fibroblasts. There was no significant difference between cell types or between sexes

(Wakayama et al. 2005b). Surprisingly, it has been shown that even differentiated neuron and lymphocyte cells can be used to establish an ntESC line (0.1%–2%), although in previous studies these cells failed to produce any cloned offspring (Hochedlinger and Jaenisch 2002; Eggan et al. 2004; Li et al. 2004; Wakayama and Yanagimachi 2001).

9.3 Application of Nuclear Transfer Embryonic Stem Cell Techniques

Embryonic stem cells derived by nuclear transfer are genetically identical to the donor and are potentially useful for therapeutic application. Therefore, therapeutic cloning may improve the treatment of neurodegenerative diseases, blood disorders, or diabetes, since therapy for these diseases is currently limited by the availability and immunocompatibility of tissue transplants (Gurdon and Colman 1999; Mombaerts 2003; Wakayama 2004). We have demonstrated that dopaminergic and serotonergic neurons can be generated from ntESCs derived from tail-tip cells (Wakayama et al. 2001). Rideout et al. (2002) reported that therapeutic cloning combined with gene therapy was able to treat a form of combined immune deficiency in mice. They made one ntESC line from an immunodeficient mutant mouse. First, the ntESC with mutated alleles was repaired by homologous recombination, thereby restoring normal gene structure. They then transplanted those ntESCs to the immunodeficient mouse. In addition, Barberi et al. (2003) reported an interesting study that showed that mouse tail or cumulus cell-derived ntESCs could be differentiated into neural cells at even higher efficiencies than fertilization-derived ESCs.

Additionally, ntES techniques can be applied to biological science. We have demonstrated that cloned mice can be generated from the nucleus of ntESCs by a second nuclear transfer (Wakayama et al. 2001). In some case, interestingly, full-term cloned mice were produced only via ntES cells but not by direct nuclear transfer from the somatic cells, or vice versa (Wakayama et al. 2005c). However, the success rate was not improved when compared to somatic cell cloning. If difficulty was experienced in the production of cloned mice from the donor cell, difficulty was also experienced while attempting to produce cloned mice by a second nuclear transfer from ntESC nuclei (Wakayama et al. 2001, 2005a). To overcome this low efficiency in the production of cloned animals, complementation with tetraploid embryos was employed, in which ntESCs were injected into the tetraploid blastocyst. As a result, almost all parts of the chimeric offspring, including germ cells, originated from ntESCs. The offspring were referred to as ES mice (Nagy et al. 1990; Wang and Jaenisch 2004). Using this technique, monoclonal mice were generated from ntESCs derived from B and T lymphocyte nuclei (Hochedlinger and Jaenisch 2002). The ntESC techniques can also be used for characterization of very rare cells in the body. If ntESCs from these rare cell nuclei are established once, the cells can proliferate infinitely. It is hypothesized that one odorant receptor gene chosen from 1,000 is controlled by DNA rearrangements in olfactory sensory neurons, such as lymphocyte nuclei. However, this could not be demonstrated because of the very low number of specific differentiated cells. Li et al. (2004) and Eggan et al. (2004) generated ntESCs from the nucleus of a single olfactory sensory neuron and demonstrated that odorant receptor gene choice is reset by nuclear transfer and is not accompanied by genomic alterations. On the other hand, ntESC techniques can be used to assess tumorigenic and developmental potential. The ntESC lines are also established from embryonal carcinoma cells (ECCs) or melanoma cell nuclei, but chimeric mice from ntESCs developed cancer with higher frequency. It has been demonstrated that nonreprogrammable genetic modifications define the tumorigenic potential (Blelloch et al. 2004; Hochedlinger et al. 2004).

9.4   Preservation of Genes from Infertile Mice Without the Use of Germ Cells

The genetically modified mouse is a powerful tool for research in the fields of medicine and biology. However, infertility was listed as a phenotype in more than half of the mutants described in one large-scale ethyl-nitrosourea (ENU) mutagenesis study (Nolan et al. 2000; Harbe de Angelis et al. 2000). This is a challenge worth taking up, as the ability to maintain such types of mutant mice as genetic resources will offer numerous advantages to research in human infertility and the biology of reproduction. Unfortunately, the success rate of somatic cell cloning is very low (Wakayama et al. 1998; Wakayama et al. 1999). Even in cases in which the cloning of a sterile mouse is successful, because of the nonreproductive nature of the phenotype, it will still be necessary to clone all subsequent generations. This represents another significant potential barrier, as the success rate of cloning from cloned mice decreases for each successive generation after the first nuclear transfer (Wakayama et al. 2000). On the other hand, ntES technology can allow us to maintain such interesting genes from mouse strains even without the use of germ cells (Wakayama et al. 2005a, 2005b).

We have found a mutant, hermaphrodite, sterile mouse in our ICR mouse breeding colony (Wakayama et al. 2005a). ICR is one of the most difficult strains to use in cloning. As a result, cloned mice have not been established. Fortunately, ntESC lines from a tail-tip fibroblast of the infertile mouse have been successfully established. Although the mutant mouse died accidentally before detail examination, the ntESC lines could be used for analysis and the Y chromosome abnormality was revealed. Since then, we have tried to make cloned mice from ntESC nuclei, but again, we could not obtain full-term offspring, probably because of the originality of this mouse. As the offspring from the ntESC lines were not obtained by conventional nuclear transfer, we have tried to make chimeric mice using either diploid embryos or tetraploid embryos. With the tetraploid complementation method, we obtained two ES mice which consisted mostly of diploid ntESCs with abnormal Y chromosome. Furthermore, they were phenotypically male and were proven infertile but not hermaphrodite. They also lacked spermatogenetic cells, although the seminiferous tubules contained Sertoli cells. In the diploid chimera, the animal with the highest contribution of ntESCs in the coat color was infertile. Interestingly, one diploid chimeric male transmitted most of its genes to the next generation via the ntESCs. Thus, ntES cells can maintain the mutant gene, but neither cloning nor tetraploid complementation chimera construction rescued the lineage of the original infertile hermaphrodite male (Wakayama et al. 2005a).

9.5 Nuclear Transfer Embryonic Stem Cell Derivation from Lethal Cloned Embryos

The success rates for the establishment of ntESC lines were much higher than those of the fully developed cloned mice. This observation was particularly noted in the case of inbred strains such as C57BL/6 and C3H/He, which are commonly used in research, but have never been successfully used to produce cloned mice (Wakayama and Yanagimachi 2001; Wakayama et al. 2001, 2005a). The establishment rate of ntESC lines is comparable to those of fertilized ESCs (Schoonjans et al. 2003). This raises the question of why the success rate for the establishment of ntESC lines is higher than that for the full-term development of a cloned mouse. On average, 30%–50% of nuclear transfer embryos develop to the blastocyst stage. However, most cloned embryos die immediately after implantation (Wakayama and Yanagimachi 1999b), which is thought to be attributable to incomplete reprogramming. This suggests that even when reprogramming is incomplete, a significant percentage of cloned embryos are able to develop to the blastocyst stage, indicating that complete reprogramming is not required for preimplantation development. If it is true that the high rates of developmental failure of cloned mice are caused by incomplete reprogramming, two possible explanations for the discrepancy between the rates of blastocyst and full-term development can be entertained. The first possibility is that the majority of ntESCs are inherently abnormal. ntESC lines are established from blastocyst-stage embryos and probably have limited gene expression requirements at this stage. Therefore, even if some embryonic genes that are essential for full-term development are not expressed, the embryos can develop to blastocysts (Boiani et al. 2002; Bortvin et al. 2003). If this is true, some ntESC lines are established from incompletely reprogrammed cloned blastocysts. Although, we demonstrated that mouse ntESCs closely resemble fertilized ESCs, using global gene expression profiles, SKY-Fish karyotype analysis and sequential nuclear transfer analysis, (Wakayama et al. 2006), it will be necessary to determine the nature and extent of such aberrations in ntESCs before considering their application in regenerative medicine. However, it is possible to use even incompletely reprogrammed ntESCs as an alternative to embryos or gametes in the preservation of genetic resources, because the abnormalities observed in cloned mice are not transmitted to their offspring (Tamashiro et al. 2002).

The second possible explanation for the different developmental competencies of cloned embryos is that the genomic reprogramming of somatic cells after nuclear transfer may occur gradually. To establish an ntESC line, the cloned blastocyst must be cultured for more than 1 month in vitro. During the culture period, survival of the cloned embryo is independent of autonomous gene expression and placental development, which in any case always shows some abnormality (Boiani et al. 2002; Bortvin et al. 2003; Tanaka et al. 2001; Ohgane et al. 2001; Inoue et al. 2002), and reprogramming is able to proceed gradually. However, in cloning, the cloned embryos must express all appropriate genes and exhibit relatively normal placental development immediately after implantation. This more stringent set of criteria may account for the higher developmental failure rates of presumably incompletely reprogrammed cloned embryos following implantation (Wakayama 2004).

9.6 Perspectives

As shown above, ntESC techniques have a great potential for use both in therapy and biological science. Recently, we have found that the reproductive cloning and ntESCs establish procedure can be improved up to 6 times by adding Trichostatin A (TSA: histone deacetylated inhibitor) (Kishigami et al. 2006), but we must examine further effect of TSA before apply for medicine. However, current methods require fresh oocytes from a healthy female. Recently, the ability of ESCs to generate gametes or interspecies oocytes shows great promise toward addressing the chronic scarcity of human oocytes for use in nuclear transfer studies. These studies can be accomplished either through the artificial creation of oocytes by in vitro differentiation (Hubner et al. 2003) or through non-human oocytes, such as cow or rabbit (Dominko et al. 1999; Chen et al. 2003), which have the potential to be reprogrammed using human somatic cell nuclei to establish human ntESC lines. Fundamental questions remain as to whether artificial oocytes have the same developmental potential as natural oocytes, and whether ntESCs and their rabbit mitochondria-containing derivatives will be tolerated by a human host's immune system on transplantation. If these doubts can be laid to rest, these techniques may one day offer an unrestricted supply of precious biological resources for use in the establishment of new ntESC lines without requiring the participation of healthy, fertile women as oocyte donors (Wakayama 2004).

References

Barberi T, Klivenyi P, Calingasan NY, Lee H, Kawamata H, Loonam K, Perrier AL, Bruses J, Rubio ME, Topf N, Tabar V, Harrison NL, Beal MF, Moore MA, Studer L (2003) Neural subtype specification of fertilization and nuclear transfer embryonic stem cells and application in parkinsonian mice. Nat Biotechnol 21:1200–1207

Blelloch RH, Hochedlinger K, Yamada Y, Brennan C, Kim M, Mintz B, Chin L, Jaenisch R (2004) Nuclear cloning of embryonal carcinoma cells. Proc Natl Acad Sci U S A 101:13985–13990

Boiani M, Eckardt S, Scholer HR, McLaughlin KJ (2002) Oct4 distribution and level in mouse clones: consequences for pluripotency. Genes Dev 16:1209–1219

Bortvin A, Eggan K, Skaletsky H, Akutsu H, Berry DL, Yanagimachi R, Page DC, Jaenisch R (2003) Incomplete reactivation of Oct4-related genes in mouse embryos cloned from somatic nuclei. Development 130:1673–1680

Chen Y, He ZX, Liu A, Wang K, Mao WW, Chu JX, Lu Y, Fang ZF, Shi YT, Yang QZ, Chen da Y, Wang MK, Li JS, Huang SL, Kong XY, Shi YZ, Wang ZQ, Xia JH, Long ZG, Xue ZG, Ding WX, Sheng HZ (2003) Embryonic stem cells generated by nuclear transfer of human somatic nuclei into rabbit oocytes. Cell Res 13:251–263

Cibelli JB, Stice SL, Golueke PJ, Kane JJ, Jerry J, Blackwell C, Ponce de Leon FA, Robl JM (1998) Transgenic bovine chimeric offspring produced from somatic cell-derived stem-like cells. Nat Biotechnol 16:642–646

Dominko T, Mitalipova M, Haley B, Beyhan Z, Memili E, McKusick B, First NL (1999) Bovine oocyte cytoplasm supports development of embryos produced by nuclear transfer of somatic cell nuclei from various mammalian species. Biol Reprod 60:1496–1502

Eggan K, Baldwin K, Tackett M, Osborne J, Gogos J, Chess A, Axel R, Jaenisch R (2004) Mice cloned from olfactory sensory neurons. Nature 428:44–49

Gurdon JB, Colman A (1999) The future of cloning. Nature 402:743–746

Hochedlinger K, Jaenisch R (2002) Monoclonal mice generated by nuclear transfer from mature B and T donor cells. Nature 415:1035–1038

Hochedlinger K, Blelloch R, Brennan C, Yamada Y, Kim M, Chin L, Jaenisch R (2004) Reprogramming of a melanoma genome by nuclear transplantation. Genes Dev 18:1875–1885

Hrabe de Angelis MH, Flaswinkel H, Fuchs H, Rathkolb B, Soewarto D, Marschall S, Heffner S, Pargent W, Wuensch K, Jung M, Reis A, Richter T, Alessandrini F, Jakob T, Fuchs E, Kolb H, Kremmer E, Schaeble K, Rollinski B, Roscher A, Peters C, Meitinger T, Strom T, Steckler T, Holsboer F, Klopstock T, Gekeler F, Schindewolf C, Jung T, Avraham K, Behrendt H, Ring J, Zimmer A, Schughart K, Pfeffer K, Wolf E, Balling R (2000) Genome-wide, large-scale production of mutant mice by ENU mutagenesis. Nat Genet 25:444–447

Hubner K, Fuhrmann G, Christenson LK, Kehler J, Reinbold R, De La Fuente R, Wood J, Strauss JF 3rd, Boiani M, Scholer HR (2003) Derivation of oocytes from mouse embryonic stem cells. Science 300:1251–1256

Inoue K, Kohda T, Lee J, Ogonuki N, Mochida K, Noguchi Y, Tanemura K, Kaneko-Ishino T, Ishino F, Ogura A (2002) Faithful expression of imprinted genes in cloned mice. Science 295:297

Inoue K, Ogonuki N, Mochida K, Yamamoto Y, Takano K, Kohda T, Ishino F, Ogura A (2003) Effects of donor cell type and genotype on the efficiency of mouse somatic cell cloning. Biol Reprod 69:1394–400

Kato Y, Tani T, Sotomaru Y, Kurokawa K, Kato J, Doguchi H, Yasue H, Tsunoda Y (1998) Eight calves cloned from somatic cells of a single adult. Science 282:2095–2098

Kawase E, Yamazaki Y, Yagi T, Yanagimachi R, Pedersen RA (2000) Mouse embryonic stem (ES) cell lines established from neuronal cell-derived cloned blastocysts. Genesis 28:156–163

Kishigami S, Mizutani E, Ohta H, Hikichi T, Thuan NV, Wakayama S, Bui HT, Wakayama T (2006) Significant improvement of mouse cloning technique by treatment with trichostatin A after somatic nuclear transfer. Biochem Biophys Res Commun 340:183–189

Li J, Ishii T, Feinstein P, Mombaerts P (2004) Odorant receptor gene choice is reset by nuclear transfer from mouse olfactory sensory neurons. Nature 428:393–399

Mombaerts P (2003) Therapeutic cloning in the mouse. Proc Natl Acad Sci U S A 100:11924–11925

Munsie MJ, Michalska AE, O'Brien CM, Trounson AO, Pera MF, Mountford PS (2000) Isolation of pluripotent embryonic stem cells from reprogrammed adult mouse somatic cell nuclei. Curr Biol 10:989–992

Nagy A, Gocza E, Diaz EM, Prideaux VR, Ivanyi E, Markkula M, Rossant J (1990) Embryonic stem cells alone are able to support fetal development in the mouse. Development 110:815–821

Nolan P, Peters J, Strivens M, Rogers D, Hagan J, Spurr N, Gray IC, Vizor L, Brooker D, Whitehill E, Washbourne R, Hough T, Greenaway S, Hewitt M, Liu X, McCormack S, Pickford K, Selley R, Wells C, Tymowska-Lalanne Z, Roby P, Glenister P, Thornton C, Thaung C, Stevenson JA, Arkell R, Mburu P, Hardisty R, Kiernan A, Erven A, Steel KP, Voegeling S, Guenet JL, Nickols C, Sadri R, Nasse M, Isaacs A, Davies K, Browne M, Fisher EM, Martin J, Rastan S, Brown SD, Hunter J (2000) A systematic, genome-wide, phenotype-driven mutagenesis programme for gene function studies in the mouse. Nat Genet 25:440–443

Ogonuki N, Inoue K, Yamamoto Y, Noguchi Y, Tanemura K, Suzuki O, Nakayama H, Doi K, Ohtomo Y, Satoh M, Nishida A, Ogura A(2002) Early death of mice cloned from somatic cells. Nat Genet 30:253–254

Ohgane J, Wakayama T, Kogo Y, Senda S, Hattori N, Tanaka S, Yanagimachi R, Shiota K (2001) DNA methylation variation in cloned mice. Genesis 30:45–50

Rideout WM 3rd, Hochedlinger K, Kyba M, Daley GQ, Jaenisch R (2002) Correction of a genetic defect by nuclear transplantation and combined cell and gene therapy. Cell 109:17–27

Schoonjans L, Kreemers V, Danloy S, Moreadith RW, Laroche Y, Collen D (2003) Improved generation of germline-competent embryonic stem cell lines from inbred mouse strains. Stem Cells 21:90–97

Tamashiro KL, Wakayama T, Blanchard RJ, Blanchard DC, Yanagimachi R (2000) Postnatal growth and behavioral development of mice cloned from adult cumulus cells. Biol Reprod 63:328–334

Tamashiro KL, Wakayama T, Yamazaki Y, Akutsu H, Woods SC, Kondo S, Yanagimachi R, Sakai RR (2002) Cloned mice have an obese phenotype not transmitted to their offspring. Nat Med 8:262–267

Tanaka S, Oda M, Toyoshima Y, Wakayama T, Tanaka M, Yoshida N, Hattori N, Ohgane J, Yanagimachi R, Shiota K (2001) Placentomegaly in cloned mouse concepti caused by expansion of the spongiotrophoblast layer. Biol Reprod 65:1813–1821

Wakayama T (2004) On the road to therapeutic cloning. Nat Biotechnol 22:399–400

Wakayama T, Yanagimachi R (1999a) Cloning of male mice from adult tail-tip cells. Nat Genet 22:127–128

Wakayama T, Yanagimachi R (1999b) Cloning the laboratory mouse. Semin Cell Dev Biol 10:253–258

Wakayama T, Yanagimachi R (2001) Mouse cloning with nucleus donor cells of different age and type. Mol Reprod Dev 58:376–383

Wakayama T, Perry ACF, Zuccotti M, Johnson KR, Yanagimachi R (1998) Full-term development of mice from enucleated oocytes injected with cumulus cell nuclei. Nature 394:369–374

Wakayama T, Shinkai Y, Tamashiro KLK, Niida H, Blanchard DC, Blanchard RJ, Ogura A, Tanemura K, Tachibana M, Perry ACF, Colgan DF, Mombaerts P, Yanagimachi R (2000) Cloning of mice to six generations. Nature 407:318–319

Wakayama T, Tabar V, Rodriguez I, Perry ACF, Studer L, Mombaerts P (2001) Differentiation of embryonic stem cell lines generated from adult somatic cells by nuclear transfer. Science 292:740–743

Wakayama S, Kishigami S, Thuan NV, Ohta H, Hikichi T, Mizutani E, Yanagimachi R, Wakayama T (2005a) Propagation of an infertile hermaphrodite mouse lacking germ cells, using nuclear transfer and embryonic stem cell technology. Proc Natl Acad Sci U S A 102:29–33

Wakayama S, Ohta H, Kishigami S, Van Thuan N, Hikichi T, Mizutani E, Miyake M, Wakayama T (2005b) Establishment of male and female nuclear transfer embryonic stem cell lines from different mouse strains and tissues. Biol Reprod 72:932–936

Wakayama S, Mizutani E, Kishigami S, Thuan NV, Ohta H, Hikichi T, Bui HT, Miyake M, Wakayama T (2005c) Mice Cloned by Nuclear Transfer from Somatic and ntES Cells Derived from the Same Individuals, J Reprod Dev 51:765–772

Wakayama S, Jakt LM, Suzuki M, Araki R, Hikichi T, Kishigami S, Ohta H, Van Thuan N, Mizutani E, Sakaide Y, Senda S, Tanaka S, Okada M, Miyake M, Abe M, Nishikawa SI, Shiota K, Wakayama T (2006) Equivalency of nuclear transfer-derived embryonic stem cells to those derived from fertilized mouse blastocysts. Stem Cells (in press)

Wang Z, Jaenisch R (2004) At most three ES cells contribute to the somatic lineages of chimeric mice and of mice produced by ES-tetraploid complementation. Dev Biol 275:192–201

10 Derivation of Germ Cells from Embryonic Stem Cells

J. Kehler, K. Hübner, H.R. Schöler

10.1 The Totipotency Cycle and the Development
of Germ Cells In Vivo and In Vitro 126
10.2 Genes Involved in the Maintenance of Pluripotency
and Primordial Germ Cell Development 127
10.3 Detection of Primordial Germ Cells In Vivo and In Vitro 129
10.3.1 Induction of Primordial Germ Cells via Bone Morphogenic
Protein Signaling in the Epiblast and in Embryoid Bodies 130
10.3.2 Expression of *Fragilis* and *Stella* in the Epiblast During
Primordial Germ Cell Specification and in Embryoid Bodies . . . 130
10.3.3 Onset of Motility in Primordial Germ Cells In Vivo and In Vitro . . 131
10.3.4 Expression of c-Kit on In Vivo-
and In Vitro-Derived Primordial Germ Cells 131
10.3.5 Expression of the Murine Vasa Homolog Postmigratory
Primordial Germ Cell Marker In Vivo and In Vitro 132
10.3.6 *SRY* Expression and Sexual Differentiation
of Primordial Germ Cells In Vivo and In Vitro 132
10.4 Meiotic Progression and Female Gamete Formation
In Vivo and In Vitro 133
10.4.1 Follicular Development In Vivo and In Vitro 134
10.4.2 Oocyte Maturation and Activation 135
10.4.3 Parthenogenetic Activation 135
10.5 Conclusions 136
References 137

Abstract. Embryonic stem cells (ESCs), derivatives of cells of early mammalian embryos, have proven to be one of the most powerful tools in developmental and stem cell biology. When injected into embryos, ESCs can contribute to tissues derived from all three germ layers and to the germline. Prior studies have successfully shown that ESCs can recapitulate features of embryonic development by spontaneously forming somatic lineages in culture. Amazingly, recently it has been shown that mouse ESCs can also give rise to primordial germ cells (PGCs) in culture that are capable of undergoing meiosis and forming both male and female gametes. While the full potential of these ES-derived germ cells and gametes remains to be demonstrated, these discoveries provide a new approach for studying reproductive biology and medicine.

10.1 The Totipotency Cycle and the Development of Germ Cells In Vivo and In Vitro

Establishment of the germline in mammals is often referred to as the totipotency cycle. Unlike in lower-order eukaryotes, the germ cell lineage has not been allocated in the embryo proper of mammals, as early epiblast cells injected into blastocysts are equipotential to enter the mouse germline (Rossant et al. 1978; Gardner and Rossant 1979). It is not until the start of gastrulation between 5.5 and 6 dpc, that cells within the proximal epiblast are competent to become PGCs (Yoshimizu et al. 2001), with the entire epiblast retaining this germline plasticity as late as 7 dpc (Tam and Zhou 1996). However, clonal analyses have demonstrated that by 7.25 dpc, a founding population of approximately 45 cells that express the tissue nonspecific isoform of alkaline phosphatase (TNAP) within the proximal epiblast have committed to becoming PGCs (Ginsburg et al. 1990; Hahnel et al. 1990; Lawson and Hage 1994). More recent genetic studies have characterized the molecular identity of this initial population further (see below).

Mouse ESCs, derivatives of inner cell mass (ICM) cells, are capable of re-entering the totipotency cycle in vivo and in vitro. When ESCs are supported by diploid or tetraploid recipient embryos, they are capable of differentiating into all three germ layers (endoderm, mesoderm, and ectoderm) and ultimately germ cells (Robertson 1986; Nagy et al. 1990; reviewed by Tam and Rossant 2003). Surprisingly, in the absence of an embryonic body plan to direct development, ESCs are capable of spontaneously recapitulating early embryonic stages culminating in the specification of the germline in culture. Three recent studies have demonstrated that ESCs can form early germ cells in both monolayer cultures and embryoid bodies (EBs) that are subsequently capable of developing into female and male gametes, respectively (Hübner et al. 2003; Toyooka et al. 2003; Geijsen et al. 2004). In addition, some of the oocyte-like structures can subsequently form structures that exhibit the appropriate polarity, morphology, and gene expression of blastocyst completing the totipotency cycle in culture (Hübner et al. 2003). While the ability of these gametes and embryonic structures to support development to term remains to be demonstrated, these in vitro assays provide a powerful new approach to test the full developmental capacity of cell lines currently accepted as pluripotential.

10.2 Genes Involved in the Maintenance of Pluripotency and Primordial Germ Cell Development

The transcription factor, Oct4, is required throughout embryonic development for the continuity of the germline, first for the maintenance of totipotential cells of the early mouse embryo and their cell line derivatives (reviewed by Pesce et al. 1998a) and then for the survival of PGCs (Kehler et al. 2004). During the totipotential, preimplantation stages of mouse embryonic development, expression of Oct4 is sustained first by maternal transcripts and protein present in fertilized oocytes and then by zygotic *Oct4* gene expression at the four- to eight-cell stage (Schöler et al. 1989; Rosner et al. 1990; Yeom et al. 1991). The coincidental downregulation of *Oct4* in the outer blastomeres of the morula during compaction and cavitation correlates with their differentiation into the trophectoderm of the preimplantation stage blastocyst, while persistent *Oct4* expression in ICM cells provides the first indication that Oct4 is required for the maintenance of pluripotency during early embryonic development (Schöler et al. 1990; Palmieri et al. 1994).

Subsequently, the requirement of Oct4 to maintain embryonic pluripotency is evident in *Oct4*-deficient peri-implantation blastocysts that consist entirely of trophectodermal cells (Nichols et al. 1998) and in ESCs where *Oct4* expression is experimentally regulated by drug administration. A 50% decrease or increase in Oct4 protein levels is sufficient to induce ESC differentiation into cells with trophoblastic or epiblastic features, respectively (Niwa et al. 2000). Around implantation (4.5 dpc), similar differential expression of *Oct4* within cells of the ICM is coincident with their subsequent differentiation into primitive endoderm and epiblast. These results suggest that the allocation of the first three lineages in the early mouse embryo depends upon defined levels of Oct4.

The continuous expression of Oct4 within the epiblast may be crucial for its maintenance between implantation and the onset of gastrulation and the ultimate formation of PGCs. During gastrulation, *Oct4* expression is progressively repressed in the primitive streak in an anterior to posterior fashion in part by the action of germ cell nuclear factor (GCNF) (Fuhrmann et al. 2001) and is finally restricted to these newly formed PGCs by 7.5 dpc (Schöler et al. 1990; Yeom et al. 1996). After gastrulation, Oct4 has a completely different but equally important function in the maintenance of the mammalian germline. Recent studies have demonstrated that Oct4 is required for the survival of early PGCs, as they undergo apoptosis after deletion of the *Oct4* locus in a conditional knock-out mouse model (Kehler et al. 2004). To date two essential but distinct functions of Oct4 during mammalian germline development have been identified: first in maintaining the totipotency of the early mouse embryo and secondly the promoting the survival of PGCs during fetal development.

The characterization of the *Oct4* regulatory elements and the development of multiple *Oct4* transgenic reporter strains of mice that differentially reflect the endogenous expression pattern of Oct4 throughout fetal development have facilitated the recent detection of germ cell production from pluripotential ESCs in culture. Initial comparisons of multiple *Oct4* reporter transgenes demonstrated that the major regulatory elements of the mouse Oct4 promoter were contained within an upstream 18-kbp genomic Oct4 fragment (*GOF18*) and that expression within totipotential ICM cells, pluripotential ESCs, and PGC required the presence of a distal element (DE), while expression in the epiblast required a proximal element (PE) (Yeom et al. 1996; Yoshimizu et al. 1999). Subsequent detection of four conserved regions (CR) within the PE and DE of the mouse, cow, and human *Oct4* orthologs (Nordhoff et al. 2001) facilitated the generation of a novel germ cell transgene (*gcOct4-eGFP*) in which part of CR3 was deleted, thereby reducing transgene activity within pluripotential ESCs while maintaining strong expression within PGCs (Hübner et al. 2003).

Additional genes have been implicated in the maintenance of pluripotency independent of Oct4 expression. Early in the study of mouse ESCs, secretion of unknown soluble factors by a mouse embryonic feeder (MEF) layer or addition of recombinant leukemia inhibitory factor (LIF) into the culture media was shown to be sufficient to support mouse ESCs in an undifferentiated state (Smith et al. 1988). Binding of LIF and other members of the interleukin-6 family of cytokines to gp-130/LIFRβ heterodimer receptors on the surface of mouse ESCs (Stahl et al. 1994; Nakamura 1998) and subsequent activation of the Jak family kinase/Stat3 signaling pathway are required for the maintenance of pluripotency in an Oct4-independent manner (Niwa et al. 1998, 2000; Matsuda et al. 1999). Independent of both these pathways, expression of *Nanog,* a divergent homeodomain gene is also required for the maintenance of pluripotency in ES and epiblast cells, while its role in PGCs has not yet been assessed (Mitsui et al. 2003; Chambers et al. 2003). In both monolayer and mouse EB cultures, removal of LIF and MEF cells that support pluripotency through activation of multiple pathways is the first step in triggering germ cell differentiation.

10.3 Detection of Primordial Germ Cells In Vivo and In Vitro

The detection of PGC formation from ESCs in vitro has been confounded by their shared expression of several cell surface receptors and transcription factors such as Oct4. Despite this limitation, the expression of several additional markers such as tissue nonspecific alkaline phosphatase (TNAP), Blimp1, stella, and fragilis indicate the onset of germ cell competence and commitment in the epiblast and can be used to detect PGC formation in vitro with some experimental finesse. Detection of acquisition of motility and activation of signaling pathways such as the SCF/c-Kit pathway that is required for subsequent PGC survival and proliferation in vivo can be used to determine whether migratory germ cells are forming in vitro. Likewise, postmigratory changes in PGCs such as expression of the *murine vasa homolog (mvh)* gene and meiotic genes such as DMC1 can also be used to identify cells committed to the germline in culture. While an early epiblast-like stage has not been directly demonstrated in either differentiating monolayer or EB cell cultures in recent studies, the rapid detection of germ cell-specific markers indicates that PGC specification occurs readily in differentiating ESC cultures.

10.3.1 Induction of Primordial Germ Cells via Bone Morphogenic Protein Signaling in the Epiblast and in Embryoid Bodies

Induction of PGCs within the proximal epiblast is dependent upon the bone morphogenic protein (BMP) signaling pathways – especially BMP4 produced by extraembryonic ectodermal (Lawson et al. 1999; Fujiwara et al. 2004) – and may be active during the formation of germ cells in culture. Genetic analyses have shown that both BMP8b and BMP2 are also required for the formation of PGCs (Ying et al. 2000; 2001) and may act in concert with BMP4 or independently (Ying et al. 2001) by activating common downstream signaling components of the SMAD-mediated pathways (Tremblay et al. 2001; Chang et al. 2001; Hayashi et al. 2002). Mixing a BMP4 overexpressing stromal cell line with germ cell-specific reporter ESC lines (discussed below) demonstrated germ cell formation as early as 1 day post-differentiation (Toyooka et al. 2003). While this approach was successful in increasing the number of PGCs formed in mixed EBs, additional studies are needed to determine whether BMP signaling pathways are active in spontaneous differentiation of ESCs into PGCs in EBs and monolayer cultures.

10.3.2 Expression of *Fragilis* and *Stella* in the Epiblast During Primordial Germ Cell Specification and in Embryoid Bodies

Further refinement of the molecular identity of the nascent PGC population within the mouse epiblast could facilitate studying germ cell development in culture. Separation of early putative PGCs from surrounding Oct4$^+$/TNAP$^+$ mesodermal cells of the epiblast by differential expression of *BMP4* and *Hoxb1*, a somatic homeobox gene, was used to identify two additional genes; *fragilis* and *stella* (also known as *Dppa3* and *Pgc7*) that distinguish the onset of germ cell competence and commitment, respectively, within the epiblast (Saitou et al. 2002). While undifferentiated mouse ESCs also express *stella* and *fragilis*, they were detected in differentiating EBs coincident with the emergence of putative germ cells between 3 and 5 days and by 7 days, respectively (Toyooka et al. 2003; Geijsen et al. 2004). More recently, the onset of expression of a transcriptional repressor, the B-lymphocyte-induced maturation protein-1 (Blimp1) was shown to be required for the formation of PGCs within the epiblast (Ohinata et al. 2005). Similar to the use of *gc-Oct4 GFP* reporters, *Blimp1 transgenes* could potentially be used to detect the specification of early germ cells from ESCs in differentiating cultures.

10.3.3   Onset of Motility in Primordial Germ Cells In Vivo and In Vitro

Acquisition of directional motility is required for germ cell colonization of the developing gonads of the mouse and provides a potential functional test to identify ES-derived PGCs in culture. In the early mouse embryo, the cluster of Blimp1/stella-positive PGCs that become allocated within the Oct4/TNAP/fragilis-positive cells of the epiblast must actively migrate from the base of the allantois to the developing genital ridges (reviewed by McLaren 2003). As PGCs enter the hindgut-endoderm and splanchnic mesoderm they exhibit directional motility toward the gonadal ridges both in vivo and in vitro over a feeder layer presumably following the same chemokine gradient (Donovan et al. 1986; Anderson et al. 2000). In co-culture experiments, ESC-derived gc-Oct4-GFP$^+$ cells were also able to home to and invade primary genital ridges in culture (Ehmcke et al, in press). The directional motility may provide a simple functional test for the identification and isolation of germ cells arising in differentiating ESC cultures.

10.3.4   Expression of c-Kit on In Vivo- and In Vitro-Derived Primordial Germ Cells

During migration, the survival and proliferation of PGCs is dependent upon their expression of the c-Kit tyrosine kinase receptor (c-Kit) and its activation by steel ligand or stem cell factor (SCF) in the developing embryonic gonad and in culture (Matsui et al. 1991; Dolci et al. 1991). Activation of the c-Kit receptor by SCF triggers the AKT kinase, mTOR (FRAP), and $p70^{S6K}$ signaling pathway in cultured PGCs (De Miguel et al. 2004). Within 3–8 days of differentiation in monolayer cultures, we routinely observe the upregulation of the c-Kit receptor on gc-Oct4-GFP$^+$ germ cells (Hübner et al. 2003). While undifferentiated ESCs also express c-Kit, these gc-Oct4-GFP$^+$/c-Kit$^+$ cells purified by flow cytometry can survive for an additional week in culture with SCF but in the absence of a feeder layer or LIF (J. Kehler and K. Hübner, unpublished data). The c-Kit receptor remains a useful functional receptor for the isolation of both in vivo and in vitro derived PGCs.

10.3.5   Expression of the Murine Vasa Homolog Postmigratory Primordial Germ Cell Marker In Vivo and In Vitro

Coincident with their arrival at the developing genital ridges, PGCs begin to express the *murine vasa homolog (mvh)* of the *Drosophila vasa* gene that encodes an RNA DEAD box helicase (Fujiwara et al. 1996; Tanaka et al. 1998), providing an exclusive marker of cells that are irreversibly committed to the germline in vivo and in vitro (reviewed by Raz 2000; Noce et al. 2001). In monolayer cultures, vasa-positive cells were first detected by IHC between 6 and 8 days of differentiation and in 1- to 3- and 5- to 7-day-old EBs (Hübner et al. 2003; Toyooka et al. 2003; Geijsen et al. 2004). Similar expression patterns were observed using *mvh*-reporter ESC lines, providing a definitive means to positively identify and isolate germ cells from EBs (Toyooka et al. 2003). A recent report of mvh-positive cells in the peripheral blood and bone marrow of mice still supports the utility of this germ cell-specific marker (Johnson et al. 2005).

10.3.6   *SRY* Expression and Sexual Differentiation of Primordial Germ Cells In Vivo and In Vitro

While the default pathway for sexual differentiation of PGCs of both genetic sexes is to adopt a female fate, the coordinated expression of the *SRY* gene within pre-Sertoli cells is required to induce the formation of a male genital ridge between 11.5 and 12.5 dpc in the mouse (reviewed by Tilmann and Capel 2002). Within the sex cords of the developing testis, the majority of karyotypically male (XY), vasa-positive PGCs will undergo mitotic arrest by 13.5 dpc and become committed to a spermatogenic fate by 14.5 dpc (Albrecht and Eicher 2001; Ohta et al. 2004). In the absence of *SRY* expression in ectopic locations such as the adrenal gland or in primary cultures, XY PGCs enter the default female pathway, adopting morphological changes consistent with oogonial differentiation and meiotic arrest (Upadhyay and Zamboni 1982; Adams and McLaren 2002). While the presence of germ cells may also influence the developing genital ridges, the concurrent development of gonadal stromal cells is critical for supporting sexual differentiation and subsequent gametogenesis.

Differences in the timing of *SRY* expression in monolayer cell cultures vs EBs can help explain the entry of PGCs derived from ESCs with an XY karyotype into female and male pathways of gametogenesis, respectively. In our monolayer cultures using differentiating XY ESCs, *SRY* expression was not detected until after the formation of female oocyte and follicle-like structures, well after initial germ cell sexual differentiation (Hübner et al. 2003), whereas, in EBs made from XY ESCs, *SRY* expression was detected by 5 days, well before the subsequent formation of haploid male germ cells capable of contributing their paternal genome to blastocysts (Geijsen et al. 2004). Likewise, when mvh-lacZ-positive cells isolated from EBs were mixed with wild-type 12.5–13.5 dpc male stromal cells expressing *SRY*, they became competent to form spermatozoa after grafting into a wild-type recipient testis (Toyooka et al. 2003). Therefore, EBs appear to have supported the development of an appropriate virilizing environment, while the misexpression of *SRY* in monolayer cultures committed emergent PGCs to adopt the default female pathway.

10.4 Meiotic Progression and Female Gamete Formation In Vivo and In Vitro

In the absence of appropriate *SRY* expression in early monolayer cultures, ESC-derived germ cells were capable of entering meiosis and forming oocyte-like structures (Hübner et al. 2003). Similar to the concurrent repression of the endogenous *Oct4* gene in female PGCs entering prophase I of meiosis by 13.5 dpc (Pesce et al. 1998b; Yoshida et al. 1998), the gc-Oct4-eGFP transgene product was downregulated in the majority of vasa-positive germ cells by day 7 of differentiation in monolayer cultures. Within 1–4 additional days, small clusters of 5–20 cells containing both somatic and vasa-positive germ cells detached from primary cultures and formed larger aggregates when cultured in suspension overnight (Hübner et al. 2003). After replating these large aggregates into in vitro maturation (IVM) media (O'Brien et al. 2003), they formed follicle-like structures containing small germ cells 10 μM in diameter.

Evidence of meiotic progression was found in both primary and secondary monolayer cultures. The mouse homolog of the yeast meiosis-specific gene, *DMC1*, is first expressed in the developing ovary at 13.5 dpc (Menke et al. 2003) and was also detectable in both primary and secondary cultures by 16 days of differentiation. In addition, follicle-like structures were observed containing oocyte-like cells between 15 and 25 μM in diameter that exhibited a range of nuclear staining patterns for the synaptonemal complex protein 3 (SCP3) similar to in vivo controls (Dobson et al. 1994), suggestive of meiotic progression concurrent with oocyte growth (Hübner et al. 2003). While the entry of PGCs into meiosis in vitro and in vivo appears to be cell-autonomous (Adams and McLaren 2002), their subsequent development into competent oocytes and ultimate meiotic arrest is dependent on the concurrent development and interaction with ovarian stromal cells.

10.4.1   Follicular Development In Vivo and In Vitro

Evidence of some of the two-way signaling interactions between oocytes and granulosa cells that are critical for primordial follicular development and the resumption of meiosis in vivo (reviewed by Gosden 2002) was also present in monolayer cultures containing growing follicle-like structures between 16 and 23 days of differentiation. In both primary and secondary cultures, we detected by quantitative RT-PCR the transient expression of the growth differentiation factor 9 (GDF9) that is normally produced by oocytes and stimulates granulosa cell proliferation (Findlay et al. 2002). During this time period, we also detected an increase in three key enzymes in the estrogenic biosynthetic pathway and in estradiol that is normally formed by granulosa cells and is required for oocyte meiotic maturation, concurrent with the growth of primary and secondary follicle-like structures containing single, central oocyte-like cells (Hübner et al. 2003).

10.4.2   Oocyte Maturation and Activation

In the mouse and most mammalian species before becoming competent to be fertilized, a fully grown oocyte must first undergo nuclear, epigenetic, and cytoplasmic maturation and resume meiosis during estrus or in culture (reviewed by Eppig et al. 2003). Progression of the reductional division stages of Meiosis I in oocytes is hallmarked by several key morphological changes, starting with the breakdown of the germinal vesicle and ending with the extrusion of first polar body (PB1) and formation of a second meiotic spindle (MII). Between 23 and 25 days of differentiation, small oocyte-like structures surrounded by a zona pellucida (ZP)-like matrix reactive for the ZP2 protein were present in both primary and secondary cultures and resembled mature MII oocytes produced from cultured primary follicles. Further staining of some of these rare oocyte-like structures with the DNA-binding dye (H33342) revealed nuclear material within a PB-like vesicle in the presumptive perivitelline space, parallel to a diffuse area within the cytoplasm, suggestive of a MII spindle (Hübner and Scholer, unpublished data). While mouse oocytes normally remain arrested at MII until after ovulation and activation by fertilization, this process can be disturbed during in vitro culture.

10.4.3   Parthenogenetic Activation

Parthenogenetic activation of MII oocytes can occur readily in vitro in most mammalian species and provides a plausible explanation for the observed development of multicellular embryonic structures in monolayer cultures. Various environmental stimuli and misexpression of key components of a meiotic cytostatic factor can permit oocyte activation without fertilization both in vitro and in vivo (reviewed by Fan and Sun 2004). In the absence of spermatozoa, multicellular (2- to 64-

cell) embryo-like, morula and blastocyst-like structures were detected floating in the supernatant and trapped within follicle-like structures in primary and secondary monolayer cultures between 42 and 45 days of differentiation. Some resembled in vivo matured blastocysts with an inner cluster of cells expressing the endogenous Oct4 protein surrounded by an outer single-cell layer expressing the trophectodermal marker TROMA by IHC (Hübner et al. 2003). While parthenogenetic activation remains a plausible mechanism for their formation, the karyotype and the developmental potential of these blastocyst-like structures remain to be determined.

10.5 Conclusions

In retrospect, the ability of mouse ESCs to spontaneously differentiate into germ cells in culture may seem obvious, as mouse ESCs can contribute to the germline in vivo. However, demonstration of the spontaneous initiation and progression of mammalian gametogenesis in culture seems utterly fantastic. This recapitulation of reproductive development in vitro has presumably been occurring in laboratories throughout the world working with mouse ESCs over the last 30 years, but remained undetected due to the difficulty of visualizing this transient process in culture. The shared expression of markers such including Oct4 may also have confounded the initial identification of PGCs arising from ESCs in vitro. While the development of new transgenes has facilitated some of the recent discoveries presented here, the challenge remains to identify genes that distinguish PGCs from ESCs.

The independent demonstrations of male and female gametogenesis in differentiated ESC cultures provides further evidence of PGC development in vitro. While the ultimate proof of function of male haploid germ cells produced directly in embryoid bodies (Geijsen et al. 2004) and of spermatozoa produced after transplantation into testes (Toyooka et al. 2003) remains to be demonstrated through the production of viable offspring, it is remarkable to see progression of this specialized differentiation process to such an advanced state. Likewise, the ability of oocyte-like structures that are capable of undergoing meiotic initiation, maturation, activation, and formation of early embryonic-like structures in monolayer cultures (Hübner et al. 2003), to support nuclear reprogramming and normal embryonic development, will require additional studies. While the capacity of both monolayer and embryoid body culture systems to support progression of oogenesis and spermatogenesis, respectively, both depend upon the simultaneous development of appropriate stromal cells, this complementary divergence should facilitate the further dissection of key signaling pathways during gametogenesis in vitro. In addition to studying reproductive development, the potential generation of limitless germ cells and gametes from ESCs provides a new approach to consider in the development of therapies in assisted reproduction.

Acknowledgements. The authors would like to acknowledge the contributions discussed in this review made by their collaborators: M. Boiani, L. Christenson, R. De La Fuente, G. Fuhrmann, R. Reinbold, J. Strauss III, and J. Wood. J. Kehler was supported by the NIH SERCA-NCRR/KO1 (RR019677–01).

References

Adams IR, McLaren A (2002) Sexually dimorphic development of mouse primordial germ cells: switching from oogenesis to spermatogenesis. Development 129:1155–1164

Albrecht KH, Eicher EM (2001) Evidence that Sry is expressed in pre-Sertoli cells and Sertoli and granulosa cells have a common precursor. Dev Biol 240:92–107

Anderson R, Copeland TK, Scholer H, Heasman J, Wylie C (2000) The onset of germ cell migration in the mouse embryo. Mech Dev 91:61–68

Beddington RS, Robertson EJ (1989) An assessment of the developmental potential of embryonic stem cells in the midgestation mouse embryo. Development 105:733–737

Burdon T, Chambers I, Stracey C, Niwa H, Smith A (1999) Signaling mechanisms regulating self-renewal and differentiation of pluripotential embryonic stem cells. Cells Tissues Organs 165:131–134

Chambers I, Colby D, Robertson M et al. (2003) Functional expression cloning of Nanog, a pluripotency sustaining factor in embryonic stem cells. Cell 113:643–655

Chang H, Matzuk MM (2001) Smad5 is required for mouse primordial germ cell development. Mech Dev 104:61–67

De Miguel MP, Cheng L, Holland EC, Federspiel MJ, Donovan PJ (2002) Dissection of the c-kit signaling pathway in mouse primordial germ cells by retroviral-mediated gene transfer. Proc Natl Acad Sci U S A 99:10458–10463

Dobson MJ, Pearlman RE, Karaiskakis A, Spyropoulos B, Moens PB (1994) Synaptonemal complex proteins: occurrence, epitope mapping and chromosome disjunction. J Cell Sci 107:2749–2760

Dolci S, Williams DE, Ernst MK et al. (1991) Requirement for mast cell growth factor for primordial germ cell survival in culture. Nature 352:809–811

Donovan PJ, Stott D, Cairns LA, Heasman J, Wylie CC (1986) Migratory and postmigratory mouse primordial germ cells behave differently in culture. Cell 44:831–838

Ehmcke J, Hübner K, Schöler HR, Schlatt S (2006) Spermatogonia: origin, physiology and prospects for conservation and manipulation of the male germline. Fertil Reprod Develop 18:7–12

Eppig J, Viveiros M, Bivens C, De La Fuente R (2003) Regulation of Mammalian oocyte maturation. In: Long P, Adashi E (eds) The ovary. Elsevier, San Diego, pp 113–129

Fan H, Sun Q (2004) Involvement of mitogen-activated protein kinase cascade during oocyte maturation and fertilization in mammals. Biol Reprod 70:535–547

Findlay JK, Drummond AE, Dyson ML et al. (2002) Recruitment and development of the follicle: the roles of the transforming growth factor-beta superfamily. Mol Cell Endocrinol 191:35–43

Fuhrmann G, Chung AC, Jackson KJ et al. (2001) Mouse germline restriction of Oct4 expression by germ cell nuclear factor. Dev Cell 1:377–387

Fujiwara T, Dunn NR, Hogan BL (2001) Bone morphogenetic protein 4 in the extraembryonic mesoderm is required for allantois development and the localization and survival of primordial germ cells in the mouse. Proc Natl Acad Sci U S A 98:13739–13744

Gardner RL, Rossant J (1979) Investigation of the fate of 4–5 day post-coitum mouse inner cell mass cells by blastocyst injection. J Embryol Exp Morphol 52:141–145

Geijsen N, Horoschak M, Kim K et al. (2004) Derivation of embryonic germ cells and male gametes from embryonic stem cells. Nature 427:148–154

Ginsburg M, Snow MH, McLaren A (1990) Primordial germ cells in the mouse embryo during gastrulation. Development 110:521–528

Gosden RG (2002) Oogenesis as a foundation for embryogenesis. Mol Cell Endocrinol 186:149–153

Hahnel AC, Rappolee DA, Millan JL et al. (1990) Two alkaline phosphatase genes are expressed during early development in the mouse embryo. Development 110:555–564

Hayashi K, Kobayashi T, Umino T et al. (2002) SMAD1 signaling is critical for initial commitment of germ cell lineage from mouse epiblast. Mech Dev 118:99–109

Hübner K, Fuhrmann G, Christenson LK et al. (2003) Derivation of oocytes from mouse embryonic stem cells. Science 300:1251–1256

Johnson J, Bagley J, Skaznik-Wikiel M, Lee HJ, Adams GB, Niikura Y, Tschudy KS, Tilly JC, Cortes ML, Forkert R, Spitzer T, Iacomini J, Scadden DT, Tilly JL (2005) Oocyte generation in adult mammalian ovaries by putative germ cells in bone marrow and peripheral blood. Cell 122:303–315

Kehler J, Tolkunova E, Koschorz B et al. (2004) Oct4 is required for primordial germ cell survival. EMBO Rep 5:1078–1083

Koshimizu U, Taga T, Watanabe M et al. (1996) Functional requirement of gp130-mediated signaling for growth and survival of mouse primordial germ cells in vitro and derivation of embryonic germ (EG) cells. Development 122:1235–12342

Lawson KA, Hage WJ (1994) Clonal analysis of the origin of primordial germ cells in the mouse. Ciba Found Symp 182:68–84; discussion 84–91

Lawson KA, Dunn NR, Roelen BA et al. (1999) Bmp4 is required for the generation of primordial germ cells in the mouse embryo. Genes Dev 13:424–436

Matsuda T, Nakamura T, Nakao K et al. (1999) STAT3 activation is sufficient to maintain an undifferentiated state of mouse embryonic stem cells. EMBO J 18:4261–4269

Matsui Y, Toksoz D, Nishikawa S, Williams D, Zsebo K, Hogan BL (1991) Effect of steel factor and leukaemia inhibitory factor on murine primordial germ cells in culture. Nature 353:750–752

McLaren A (2003) Primordial germ cells in the mouse. Dev Biol 262:1–15

Menke DB, Koubova J, Page DC (2003) Sexual differentiation of germ cells in XX mouse gonads occurs in an anterior-to-posterior wave. Dev Biol 262:303–312

Mitsui K, Tokuzawa Y, Itoh H et al. (2003) The homeoprotein Nanog is required for maintenance of pluripotency in mouse epiblast and ES cells. Cell 113:631–642

Nagy A, Gocza E, Diaz EM et al. (1990) Embryonic stem cells alone are able to support fetal development in the mouse. Development 110:815–821

Nagy A, Rossant J, Nagy R, Abramow-Newerly W, Roder JC (1993) Derivation of completely cell culture-derived mice from early-passage embryonic stem cells. Proc Natl Acad Sci U S A 90:8424–8428

Nakamura T, Arai T, Takagi M et al. (1998) A selective switch-on system for self-renewal of embryonic stem cells using chimeric cytokine receptors. Biochem Biophys Res Commun 248:22–27

Nichols J, Zevnik B, Anastassiadis K et al. (1998) Formation of pluripotential stem cells in the mammalian embryo depends on the POU transcription factor Oct4. Cell 95:379–391

Niwa H, Burdon T, Chambers I, Smith A (1998) Self-renewal of pluripotential embryonic stem cells is mediated via activation of STAT3. Genes Dev 12:2048–2060

Niwa H, Miyazaki J, Smith AG (2000) Quantitative expression of Oct-3/4 defines differentiation, dedifferentiation or self-renewal of ES cells. Nat Genet 24:372–376

Noce T, Okamoto-Ito S, Tsunekawa N (2001) Vasa homolog genes in mammalian germ cell development. Cell Struct Funct 26:131–136

Nordhoff V, Hübner K, Bauer A et al. (2001) Comparative analysis of human, bovine, and murine Oct-4 upstream promoter sequences. Mamm Genome 12:309–317

O'Brien MJ, Pendola JK, Eppig JJ (2003) A revised protocol for in vitro development of mouse oocytes from primordial follicles dramatically improves their developmental competence. Biol Reprod 68:1682–1686

Ohinata Y, Payer B, O'Carrol D et al. (2005) Blimp1 is a critical determinant of the germ cell lineage in mice. Nature 436:207–213

Ohta H, Wakayama T, Nishimune Y (2004) Commitment of fetal male germ cells to spermatogonial stem cells during mouse embryonic development. Biol Reprod 70:1286–1291

Palmieri SL, Peter W, Hess H, Schöler HR (1994) Oct-4 transcription factor is differentially expressed in the mouse embryo during establishment of the first two extraembryonic cell lineages involved in implantation. Dev Biol 166:259–267

Pesce M, Gross MK, Schöler HR (1998a) In line with our ancestors: Oct-4 and the mammalian germ. Bioessays 20:722–732

Pesce M, Wang X, Wolgemuth DJ, Schöler H (1998b) Differential expression of the Oct-4 transcription factor during mouse germ cell differentiation. Mech Dev 71:89–98

Raz E (2000) The function and regulation of vasa-like genes in germ-cell development. Genome Biol 1:Rev. 1017

Robertson EJ (1986) Pluripotential stem cell lines as a route into the mouse germline. Trends Genet 2:9–13

Rosner MH, Vigano MA, Ozato K et al. (1990) A POU-domain transcription factor in early stem cells and germ cells of the mammalian embryo. Nature 345:686–692

Rossant J, Gardner RL, Alexandre HL (1978) Investigation of the potency of cells from the postimplantation mouse embryo by blastocyst injection: a preliminary report. J Embryol Exp Morphol 48:239–247

Saitou M, Barton SC, Surani MA (2002) A molecular programme for the specification of germ cell fate in mice. Nature 418:293–300

Schöler HR, Hatzopoulos AK, Balling R, Suzuki N, Gruss P (1989) A family of octamer-specific proteins present during mouse embryogenesis: evidence for germline-specific expression of an Oct factor. EMBO J 8:2543–2550

Schöler HR, Ruppert S, Suzuki N, Chowdhury K, Gruss P (1990) New type of POU domain in germline-specific protein Oct-4. Nature 344:435–439

Smith AG, Heath JK, Donaldson DD et al. (1988) Inhibition of pluripotential embryonic stem cell differentiation by purified polypeptides. Nature 336:688–690

Stahl N, Boulton TG, Farruggella T et al. (1994) Association and activation of Jak-Tyk kinases by CNTF-LIF-OSM-IL-6 beta receptor components. Science 263:92–95

Tam PL, Rossant J (2003) Mouse embryonic chimeras: tools for studying mammalian development. Development 130:6155–6163

Tam PP, Zhou SX (1996) The allocation of epiblast cells to ectodermal and germ-line lineages is influenced by the position of the cells in the gastrulating mouse embryo. Dev Biol 178:124–132

Tanaka SS, Toyooka Y, Akasu R et al. (2000) The mouse homolog of Drosophila Vasa is required for the development of male germ cells. Genes Dev 14:841–853

Tilmann C, Capel B (2002) Cellular and molecular pathways regulating mammalian sex determination. Recent Prog Horm Res 57:1–18

Toyooka Y, Tsunekawa N, Takahashi Y et al. (2000) Expression and intracellular localization of mouse Vasa-homologue protein during germ cell development. Mech Dev 93:139–149

Toyooka Y, Tsunekawa N, Akasu R, Noce T (2003) Embryonic stem cells can form germ cells in vitro. Proc Natl Acad Sci U S A 100:11457–11462

Tremblay KD, Dunn NR, Robertson EJ (2001) Mouse embryos lacking Smad1 signals display defects in extra-embryonic tissues and germ cell formation. Development 128:3609–3621

Upadhyay S, Zamboni L (1982) Ectopic germ cells: natural model for the study of germ cell sexual differentiation. Proc Natl Acad Sci U S A 79:6584–6588

Yeom YI, Ha HS, Balling R, Schöler HR, Artzt K (1991) Structure, expression and chromosomal location of the Oct-4 gene. Mech Dev 35:171–179

Yeom YI, Fuhrmann G, Ovitt CE et al. (1996) Germline regulatory element of Oct-4 specific for the totipotential cycle of embryonal cells. Development 122:881–894

Ying Y, Liu XM, Marble A, Lawson KA, Zhao GQ (2000) Requirement of Bmp8b for the generation of primordial germ cells in the mouse. Mol Endocrinol 14:1053–1063

Ying Y, Qi X, Zhao GQ (2001) Induction of primordial germ cells from murine epiblasts by synergistic action of BMP4 and BMP8B signaling pathways. Proc Natl Acad Sci U S A 98:7858–7862

Yoshida K, Kondoh G, Matsuda Y et al. (1998) The mouse RecA-like gene Dmc1 is required for homologous chromosome synapsis during meiosis. Mol Cell 1:707–718

Yoshimizu T, Sugiyama N, De Felice M et al. (1999) Germline-specific expression of the Oct-4/green fluorescent protein (GFP) transgene in mice. Dev Growth Differ 41:675–684

Yoshimizu T, Obinata M, Matsui Y (2001) Stage-specific tissue and cell interactions play key roles in mouse germ cell specification. Development 128:481–490

11 Germline Recruitment in Mice: A Genetic Program for Epigenetic Reprogramming

Y. Ohinata, Y. Seki, B. Payer, D. O'Carroll, M.A. Surani, M. Saitou

11.1 Introduction 144
11.2 Key Epigenetic Modifications During Mammalian Development . 145
11.3 Specification of Germ Cell Fate in Mice 147
11.3.1 Early Development and Emergence of the New Generation
of Germline Cells in Mice 147
11.3.2 Origin and Properties of the Germ Cell Lineage 150
11.4 Epigenetic Reprogramming in Early Germ Cells 157
11.4.1 Epigenetic Reprogramming
in Migrating Primordial Germ Cells in Mice 157
11.4.2 Epigenetic States in Early Germ Cells in *Caenorhabditis elegans*
and *Drosophila melanogaster* 163
11.5 Conclusion and Perspectives 164
References 167

Abstract. Germ cells provide an enduring link between generations and therefore must possess the fundamental ability of reprogramming their genome to generate a totipotent state. We wish to understand the molecular basis of the unique properties of the mammalian germ line. Recently we identified Blimp1, a potent transcriptional repressor of a histone methyltransferase subfamily, as a critical determinant of the germ cell lineage in mice. Surprisingly, Blimp1 expression marks the origin of the germ line in proximal epiblast cells in pregastrulation embryos, substantially earlier than previously thought. Furthermore, we showed that established primordial germ cells undergo extensive erasure of genome-wide histone H3 lysine 9 dimethylation (H3K9me2) and DNA methylation, two major repressive epigenetic modifications, and instead acquire high levels of H3-K27 trimethylation (H3K27me3) in their migration period. We suggest that germline specification is a genetic system for the orderly reprogramming of the cells' epigenome toward a totipotent state, with reacquisition of totipotency-associated transcription factors and continued Blimp1 expression preventing their reversion to an explicit pluripotent state or somatic differentiation.

11.1   Introduction

Germ cells differ from somatic cells in their unique and fundamental necessity to transmit and propagate the genetic information to subsequent generations. These cells exhibit a remarkably complex remodeling of their genome throughout their development, including epigenetic reprogramming and meiosis, while maintaining an underlying totipotency (Li 2002; Surani 2001). They also give rise to stem cells, known as germline stem cells (Brinster 2002; Lin 1997), in the both sexes, which maintain a supply of haploid gametes for the generation of new organisms upon their fusion at fertilization. A detailed understanding of the molecular mechanisms governing germline development will provide essential knowledge regarding genome totipotency, epigenetic mechanisms controlling cellular plasticity, and stem cell development.

Specification of germ cell fate is an essential event during the life cycle of the organism, which segregates germ cells that will eventually generate the totipotent state, from somatic lineages that progress toward terminally differentiated states. Germ cell fate is specified by one of two distinct mechanisms in different organisms: preformation or epigenesis. Preformation or predetermined germ cell specification involves inheritance of a specific cytoplasmic structure, the germ plasm, by specific blastomeres during early cleavage divisions of the zygote and early embryos. The germ plasm confers germ cell fate in cells that inherit this structure. Preformation is seen in many organisms including *Caenorhabditis elegans*, *Drosophila melanogaster*, *Xenopus laevis*, and *Danio rerio* (zebrafish) (Eddy 1975; Wylie 1999). Components of the germ plasm induce a robust global transcriptional repression that allows germ cells to escape from somatic cell differentiation and fate during the earliest stages of development (Blackwell 2004; Leatherman and Jongens 2003). The molecules and mechanisms that accomplish this differ widely between organisms (see Sect. 11.4.2).

In contrast, in epigenesis or regulated germ cell formation, germ cell fate is imposed on a subset of potentially equivalent cells by specific inductive signals from adjacent tissues (Lawson et al. 1999; Lawson and Hage 1994; McLaren 1999; Saitou et al. 2003). In mammals and several other organisms, germ cell fate is determined by epigenesis, and this mechanism appears to be the dominant mode more ancestral to the Metazoa (Extavour and Akam 2003; Johnson et al. 2003). Studies on the early mouse germ line have demonstrated the importance of epigenetic reprogramming including erasure and re-establishment of parental imprints (Hajkova et al. 2002; Lee et al. 2002; Surani 2001). Furthermore, studies have shown that it is possible to derive pluripotent stem cells called embryonic germ cells (EGCs) from primordial germ cells (PGCs) (Matsui et al. 1992). EGCs are similar to pluripotent embryonic stem cells (ESC), demonstrating the underlying pluripotency of PGCs, the common precursors of oocytes and sperm. However, the precise mechanisms that control germ cell fate in mammals and segregate these cells from somatic cells has remained unclear until relatively recently.

Identifying the mechanisms regulating germ cell determination and differentiation in mice will provide information of the molecular basis by which totipotency may be reacquired/retained as well as extensive epigenetic reprogramming that occurs in germ cells. We present our recent studies on the molecular basis of germ cell segregation and the extensive epigenetic reprogramming of the germ cell lineage in mice, and compare these events with those in other model organisms where germ cell specification involves inheritance of preformed germ cell determinants.

11.2   Key Epigenetic Modifications During Mammalian Development

Methylation of cytosine within the CpG dinucleotide motif is a key epigenetic modification, which is functionally associated with parental imprinting, gene expression, and genome structure (Bird 2002). Generally, methylated DNA acts as a repressive element for transcription by creating a local heterochromatic structure. Methylated cytosine residues are distributed throughout the genome and are particularly concentrated at pericentric heterochromatin.

Covalent modification of specific residues in the N-terminal tails of histones is another major epigenetic modification critical to the control of both gene expression and genome structure (Jenuwein and Allis 2001; Lachner and Jenuwein 2002; Lachner et al. 2003). In particular, methylation of lysine (K) residues of histone H3, in mono-, di-, or trimethylated states, is essential for the generation of distinct chromatin domains (Peters et al. 2003). Among the distinct methylated residues of H3, K9 and K27 have been well characterized. A family of enzymes possessing SET (Suv39h [Suppressor of variegation 39h], Enhancer of zeste, Trithorax) domains is responsible for the observed histone methyltransferase (HMTase) activity. G9a and Glp are examples of HMTases directing H3K9 mono- and dimethylation (H3K9me and H3K9me2, respectively) in euchromatin, thereby repressing transcription (Peters et al. 2003; Tachibana et al. 2002, 2005). All three H3K27 methylation states are broadly present in euchromatin as well. In facultative heterochromatin including inactive X chromosomes, a large amount of H3K27me3 by the Enhancer of zeste (Ezh2)/Embryonic ectoderm development (Eed) complex is detected, and H3K9me2 is also present (Erhardt et al. 2003; Peters et al. 2003; Plath et al. 2003; Silva et al. 2003). In pericentric heterochromatin, Suv39h-dependent H3K9me3 and Suv39h-independent H3K27me predominate (Peters et al. 2001, 2003). In contrast to these repressive modifications, H3-K4 methylation mediated by SET7/9 (Wang et al. 2001) and H3-K9 acetylation mediated by histone acetyltransferases are typically coupled to transcriptionally permissive/active euchromatin, and these two modifications mutually exclude H3-K9 methylation (Jenuwein and Allis 2001; Lachner and Jenuwein 2002; Lachner et al. 2003; Wang et al. 2001).

11.3 Specification of Germ Cell Fate in Mice

11.3.1 Early Development and Emergence
of the New Generation of Germline Cells in Mice

Development begins with the fusion of egg and sperm, the terminal and most specialized cells of the germ cell lineage (Fig. 1). The mouse oocyte inherits essential factors for the initiation of development and early reprogramming of the parental genomes, especially the male genome that is packaged in protamines (Morgan et al. 2005). This reprogramming includes rapid replacement of protamines with maternally deposited histones followed by active DNA demethylation, a process that is complete before the onset of DNA replication in the paternal pronucleus. Some regions of the sperm genome are resistant to demethylation, including those of paternally imprinted genes (Olek and Walter 1997), IAP retrotransposons (Lane et al. 2003), and pericentric heterochromatin (Rougier et al. 1998; Santos et al. 2002). The mechanisms regulating the active DNA demethylation of the sperm genome are unknown despite its potential relevance for the epigenetic control of cell fate, but both direct and indirect demethylation have been proposed as being responsible (Morgan et al. 2005). Interestingly, demethylation of the paternal genome is conserved across several mammalian species (Dean et al. 2001), suggesting that this event has functional significance. The histones of the paternal and maternal genomes are asymmetrically modified (Arney et al. 2002; Santos et al. 2005), and this asymmetry seems to contribute to the DNA demethylation only for the paternal genome. Resetting the epigenetic modifications for the initiation of embryonic development is apparently a well-programmed process (Santos et al. 2005), as is the case for genome reprogramming in early germ cells (see Sect. 11.4.1).

Zygotic transcription starts as early as the late one-cell stage (minor transcription), which becomes progressively very active (major transcription) at the two-cell stage (Hamatani et al. 2004). The first overt differentiation event during mammalian development is the formation of the trophectoderm, the precursor cells of the placenta, which occurs at the outer surface of preimplantation embryos at the late eight-cell stage, and the transcription factor Cdx2 is apparently essential for this process (Strumpf et al. 2005). At embryonic day (E) 3.5, embryos form

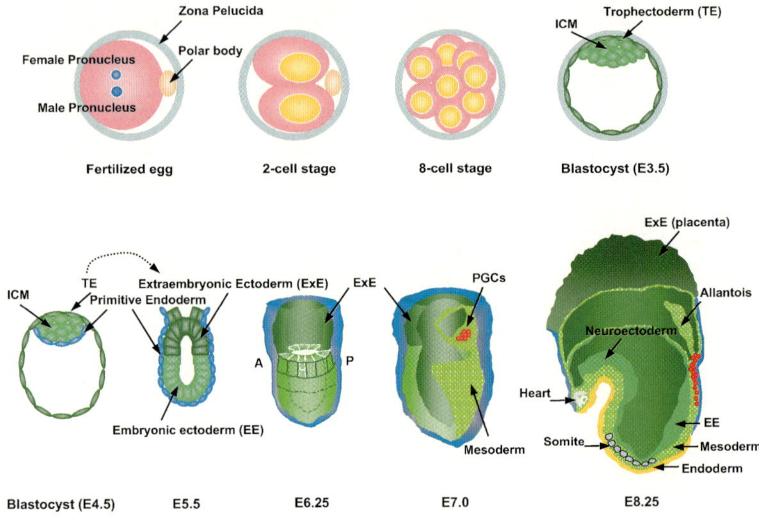

Fig. 1. Schematic representation of mouse early development and the emergence of the germline. See text for details

a structure called the blastocyst with the inner cell mass (ICM) inside, which contains pluripotent epiblast cells. ESCs are derived in culture from the ICM (Evans and Kaufman 1981; Martin 1981). The formation of this pluripotent lineage critically depends on transcription factors such as Oct4 (Nichols et al. 1998), nanog (Mitsui et al. 2003), and Sox2 (Avilion et al. 2003). Epigenetically, these ICM cells have the lowest levels of DNA methylation, which is the result of replication-dependent passive DNA demethylation from the two-cell stage onward (Li 2002; Rougier et al. 1998). In contrast, the level of H3K27me3, mediated by the Ezh2/Eed complex, is high in these cells (Erhardt et al. 2003) (see Sect. 11.4.1). This modification is essential for early development and for the derivation of ESCs (Erhardt et al. 2003; O'Carroll et al. 2001), suggesting that H3K27me3 is important for maintaining the epigenetic plasticity of pluripotent epiblast and ESCs.

ICM cells are the starting point of cellular differentiation during mammalian development, but prior to this stage is a period of reprogramming of the parental germline genomes required for proper embryogenesis. Global levels of DNA methylation increase hereafter with the activity of de novo DNA methyltransferases Dnmt3a and Dnmt3b (Li 2002; Okano et al. 1999). Additionally, the histone methyltransferases G9a (Tachibana et al. 2002), Glp (Tachibana et al. 2005), and Eset (Dodge et al. 2004), among others, are also essential for early embryogenesis. At E4.5, blastocysts implant with the trophectoderm and ICM cells differentiate into either primitive endoderm or pluripotent epiblast cells, which form a cup-shaped epithelial sheet at E5.5. It is from this population of cells that all the somatic cells as well as the next generation of germ cells arise (Gardner et al. 1985; Gardner and Rossant 1979). It is of considerable interest to understand when, where, and how the new germ cell lineage arises within this apparently functionally equivalent epiblast cells, and how they acquire their distinctive cellular and genomic characteristics.

Histologically, it is at E7.5 (early bud [EB] stage) during gastrulation when a cluster of approximately 40 primordial germ cells (PGCs) is first identified with their characteristic alkaline phosphatase activity within the extraembryonic mesodermal cells at the base of the incipient allantois (Chiquoine 1954; Ginsburg et al. 1990). When examined by electron microscopy, these cells exhibit a distinct electron-dense cytoplasm, reflecting an already significantly different molecular composition (Saitou et al. 2002, 2003). Clonal analysis of the fate of E6.0 (prestreak [PS] stage) to E6.5 (early-streak [ES] stage) epiblast cells showed that precursor cells that contribute to PGCs reside within the proximal part of the epiblast cells (Lawson and Hage 1994). Importantly, this analysis showed that PGC fate is not restricted in these precursors, since the clonal descendants of any single cells contributing to PGCs include cells in somatic lineages, most notably the allantois mesoderm. Additionally, signaling molecules including bone morphogenetic proteins, Bmp4 and Bmp8b, both originating from the extraembryonic ectoderm, are required for the formation of PGCs (Lawson et al. 1999; Ying et al. 2000). Furthermore, signal transducers of these molecules, including the receptor Alk2 (de Sousa Lopes et al. 2004), Smad1 (Hayashi et al. 2002; Tremblay et al. 2001), Smad4 (Chu et al. 2004), and Smad5 (Chang and Matzuk 2001), are also essential for PGC specification. However, mutant embryos deficient in Bmp4, Bmp8b, Alk2, and Smad1 exhibit abnormalities in allantois formation as well, leading to the hypothesis that these signaling pathways are necessary for the priming of the proximal epiblast cells toward germ cell fate and extraembryonic mesoderm. However, the precise role of these signaling molecules in germline segregation remains unclear.

11.3.2   Origin and Properties of the Germ Cell Lineage

Single-Cell Analysis for the Specification of Germ Cell Fate

To identify the origin and properties of the germ cell lineage, we set out to systematically isolate molecules specifically involved in germ cell specification. However, as the number of founder PGCs in the extraembryonic mesoderm is small, and as they are intimately surrounded by somatic neighbors, we decided to create single-cell cDNA libraries originated from founder PGCs at E7.5 (EB stage) (Saitou et al. 2002). A small embryonic fragment of approximately 250 cells with founder PGCs was cut out and dissociated into single cells to derive single-cell cDNAs by a global amplification of mRNAs expressed in the original single cells. The origin of the resultant single cell cDNAs was retrospectively identified by the expression of a set of positive and negative markers. We originally used *Tissue nonspecific alkaline phosphatase* (*Tnap*) as a positive marker and *Bmp4* and *Hoxb1* as negative markers for the identification of putative founder PGCs. The key transcription factor for pluripotency, *Oct4*, is expressed similarly in almost all the cells we screened. The cDNAs from a single putative founder PGCs were differentially screened with the cDNAs of somatic cell origin to identify molecules specifically represented in founder PGCs. This screen initially identified two genes, *fragilis* and *stella*, highly and specifically expressed in PGCs, respectively. *fragilis* is a novel member of interferon-inducible transmembrane protein family, thought to be involved in cell aggregation and cell cycle elongation (Deblandre et al. 1995; Evans et al. 1990, 1993). In contrast, *stella* is a small nucleocytoplasmic protein whose expression is restricted to preimplantation embryos and oocytes (Sato et al. 2002).

Precise in situ hybridization analysis and single-cell cDNA expression profiling predicted a molecular program for the specification of germ cell fate in mice (Saitou et al. 2002). *fragilis* expression is concentrated in the proximal epiblast cells at E6.25–E6.5 (not-streak [0S]/ES stage), and it is dependent on Bmp4. Expression of *fragilis* marks competence for germ cell formation. These cells move posteriorly where *fragilis* expression intensifies further and the expression of *stella* begins around E7.0 (late-streak [LS] stage/not-bud [0B] stage) specifically in the cells expressing the highest level of *fragilis*. Additionally, expression of the *Hox* genes, *Hoxb1* and *Hoxa1*, which are highly upregulated in somatic neighbors, are specifically repressed in these cells. We propose that these *stella*-positive and *Hox*-negative cells are the lineage-restricted founder PGCs, and the repression of *Hox* gene expression is fundamental for these cells to escape from the somatic fate. We also note that although *Oct4* is expressed similarly both in the PGCs and in their somatic neighbors, its partner *Sox2* is confined to PGCs (Saitou et al., unpublished data), indicating that the Oct4-Sox2 complex is only present in PGCs; this difference could be critical for the underlying pluripotency in PGCs. Since the expression of *stella* and the repression of *Hox* genes are highly correlated, *stella* was thought to be a candidate molecule regulating this process. However, germ cell fate specification itself was not affected by the absence of stella, as shown by knock-out studies (Payer et al. 2003). Interestingly, stella-deficient oocytes, however, exhibited a severely reduced developmental potential when fertilized, revealing that *stella* is a novel maternal effect gene in mice (Payer et al. 2003), although the mechanism of stella function is currently unknown.

Blimp1 Is a Critical Determinant of the Germ Cell Lineage in Mice

To identify additional key molecules involved in the specification of germ cell fate, we further analyzed single germ cell cDNAs, which included examination of differential expression of candidate genes, such as histone methyltransferases and polycomb genes. These studies showed that *B-lymphocyte maturation-induced protein-1* (*Blimp1*) is detectable exclusively in germ cell cDNAs (Ohinata et al. 2005). *Blimp1* is a transcriptional repressor with a N-terminal PR domain (a derivative of the SET domain), followed by a proline-rich region, five C2H2 zinc fingers, and a C-terminal acidic domain. *Blimp1* was originally identified as a gene specifically induced upon B-cell line differentiation into plasma cells (Turner et al. 1994). During development, Blimp-1 is widely expressed after E7.5, including migrating germ cells (Chang et al. 2002), and therefore we decided to analyze this gene further.

Blimp1 Marks the Origin of the Germ Cell Lineage   We first carefully analyzed the expression of *Blimp1* in early embryos (Ohinata et al. 2005) (Fig. 2). Consistent with the single-cell analysis, in situ hybridization of LS stage (E7.0) embryos revealed expression of *Blimp1* in the proximal-posterior extraembryonic mesodermal region, the area of founder PGC formation, in addition to expression in the visceral endoderm, as reported previously. Because *Blimp1* expression appeared more specific and earlier than that of *fragilis* and *stella*, respectively, we set out to determine the onset of *Blimp1* expression in earlier embryos. Surprisingly, we found that in the embryo proper, *Blimp1* expression begins in the single layer of the most proximal epiblast cells in the 0S stage embryos (E6.25) at the onset of gastrulation. Moreover, the earliest *Blimp1* expression was restricted to four to eight cells at one side (posterior) of the epiblast (one end of the short embryonic axis). These findings suggest that germ cell fate restriction may occur much earlier than previously thought in a restricted population of *Blimp1*-positive cells. Interestingly, *Blimp1* is not expressed in ICM cells or ES cells, further suggesting that it may have a specific role in the specification of germ cells.

To visualize and examine *Blimp1* expression more precisely, we generated transgenic lines expressing membrane-targeted EGFP under the control of the *Blimp1* upstream element (*Blimp1*-mEGFP) using bacterial artificial chromosome (BAC) modification. These transgenic lines recapitulated precisely the expression of *Blimp1* (Fig. 2). The four to eight *Blimp1*-positive cells emerge in the most proximal layer of the epiblast that are in direct contact with the extraembryonic ectoderm at E6.25–6.5. These cells increase in number and form a tight cluster at the mid-streak (MS) stage (E6.75) with approximately 20 cells in direct contact with the most proximal epiblast cells. The number of *Blimp1*-positive cells at the EB stage (E7.5) was on average 40, and nearly all of these cells exhibited alkaline phosphatase activity, a classic marker of the founder PGCs. Taken together, these data strongly suggest that the initial

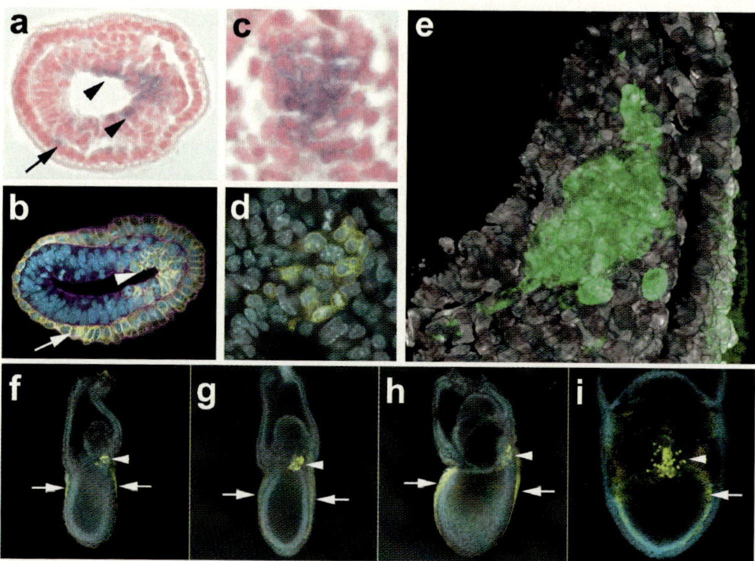

Fig. 2. a–i *Blimp1* expression in gastrulating mouse embryos. **a** Transverse view of *Blimp1* expression in the epiblast (*arrowhead*) and visceral endoderm (*arrow*) by in situ hybridization at the MS stage. Anterior is to the left. **b** *Blimp1* expression (*yellow*) visualized by the *Blimp1*-mEGFP reporter at the E/MS stage. This embryo was counterstained with DAPI (*blue*) and phalloidine (*purple*). **c, d** Posterior view of *Blimp1* expression at the LS stage. **e** Projected three-dimensional image of confocal sections of *Blimp1* expression (*green*) at the EB stage. **f–i** *Blimp1* expression in PGCs (*arrowheads*) and in visceral endoderm (*arrows*) in the LS (**f**), 0B (**g**), EB (**h**), and M/LB (**i**) stage embryos. **f–h** Lateral views and **i** posterior view

Blimp1-positive epiblast cells give rise to the germ cell lineage. To further reinforce this conclusion, mice expressing Cre recombinase under the control of the *Blimp1* upstream element (*Blimp1-Cre*) were crossed with mice expressing GFP under the control of a constitutively active promoter following excision of a loxP-flanked stop codon (*ROSA26-EGFP^f*). Theoretically all cells expressing *Blimp1* at any point during development should be permanently labeled with GFP. Remarkably, we found that almost all the cells expressing GFP in this cross were positive for *stella* at the LB stage (E7.75), demonstrating that cells expressing *Blimp1* are the lineage-restricted germ cell precursors that subsequently go on to express *stella*.

These data contrast with the alternative hypothesis proposed previously based on clonal analysis that predicts that germ cell fate is not restricted in the epiblast (Lawson and Hage 1994). Careful examination of the localization of *Blimp1*-positive cells in the OS and ES stage embryos revealed that these cells do reside in the epiblast with some *Blimp1*-negative nascent mesodermal cells even after the onset of gastrulation (Ohinata et al. 2005) (Fig. 2a, b). It is possible that the clonal analysis may have failed to label the low numbers of very early *Blimp1*-positive epiblast cells. The precise mechanism controlling the overall number of PGCs at the EB stage (40) remains unknown, however. It is possible that the epiblast clones contributing to PGCs in the clonal analysis were initially negative for *Blimp1* and divided to form one *Blimp1*-positive cell and one somatic descendant. In order to resolve these issues, it will be important to precisely determine the cell cycle state and the accretion of the *Blimp1*-positive early germ cell precursors.

Critical Role of Blimp1 in the Establishment of Primordial Germ Cells   The gene knock-out study clearly showed that *Blimp1* function in the epiblast is required for the proper formation of authentic germ cells (Ohinata et al. 2005; Vincent et al. 2005). At E8.5, the gross appearance of *Blimp1*-deficient embryos was normal, including a normal allantois. However, when stained for alkaline phosphatase activity, virtually no migrating PGCs were seen in these embryos. At the EB stage, wild-type embryos contain approximately 40 alkaline phosphatase-positive founder PGCs as a cluster, with some cells beginning to migrate, and, at the LB stage, many of the approximately 60 PGCs have migrated and dispersed throughout the embryo. In contrast, in the *Blimp1*-deficient embryos, we observed a cluster of only 20 alkaline phosphatase-positive cells at the EB stage. Remarkably, even at the LB stage, the size of the cluster remained quite small, with little or no cell migration (Ohinata et al. 2005). These findings demonstrate that *Blimp1* is necessary at a very early stage (earlier than the EB stage) in germ cell specification. When examined by single cell analysis, *Blimp1*-deficient cells did not consistently repress *Hox* gene expression, a condition that occurs in wild-type embryos. Additionally, expression of several genes normally upregulated in germ cells was not detected in these mutant cells. These findings clearly indicate that Blimp1 is a key transcriptional regulator of germ cell lineage development (Fig. 3).

Overexpression of Blimp1 in a B-cell line is sufficient to drive plasma cell differentiation (Turner et al. 1994), and remarkably, this event is associated with the repression of numerous key molecules associated with a mature B-cell phenotype (Shaffer et al. 2002). Moreover, conditional gene knock-out of *Blimp1* in the B-cell lineage disrupts plasma cell differentiation (Shapiro-Shelef et al. 2003). Therefore, Blimp1 is regarded

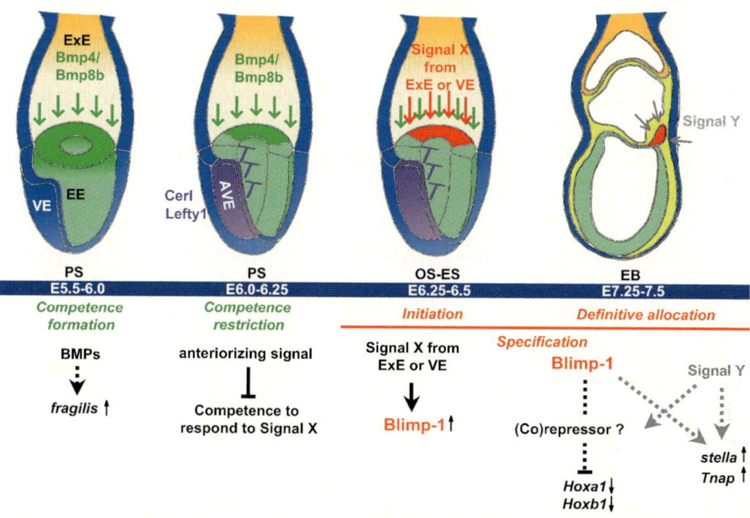

Fig. 3. A proposed model for the specification of germ cell fate in mice. Proximal epiblast cells are exposed to signaling molecules such as Bmp4 and Bmp8b from extraembryonic ectoderm (*ExE*), becoming competent for germ cell formation. However, anteriorizing signals from anterior visceral endoderm (*AVE*) may inhibit this competence in anterior epiblast cells. Eventually, only a few of the most proximal posterior epiblast cells begin to express *Blimp1*. This may be triggered by an unknown signal X from ExE or VE. *Blimp1*-expressing cells may need another signal Y to fully differentiate *stella*-positive founder PGCs. See text for details

as a master regulator of plasma cell differentiation. Mechanistically, Blimp1 functions as a transcriptional repressor by recruiting a repressor complex including Groucho and HDAC2 through its proline-rich region (Ren et al. 1999), and the histone methyltransferase G9a, essential for H3K9me2, by its zinc fingers (Gyory et al. 2004). No enzymatic activity has yet been found for the SET domain in Blimp1. Blimp1 can also activate the expression of some genes with its carboxyl terminal acidic domain (Sciammas and Davis 2004). Therefore, Blimp1 can potentially exert diverse functions in a context-dependent manner.

Consistent with this wide array of functions, Blimp1 has recently been shown to play important roles in many places where it is expressed across several vertebrate species. In zebrafish, Blimp1 activity is required for the lineage specification of at least three cell types. A hypomorphic allele of *u-boot* (*ubo*), the zebrafish ortholog of *Blimp1*, results in the loss of slow-twitch muscle fibers (Baxendale et al. 2004). In this mutant, muscle precursors fail to differentiate into slow-twitch fibers in response to Shh. Ubo activates expression of the slow myosin heavy chain and represses expression of the fast myosin heavy chain. Ubo also acts in the specification of the common progenitors for the neural crest and sensory neurons downstream of Bmp signaling (Hernandez-Lagunas et al. 2005; Roy and Ng 2004). In *Xenopus* and zebrafish, *Blimp1* is also expressed in the anterior mesoendoderm and prechordal plate, signaling centers for forebrain patterning (de Souza et al. 1999; Wilm and Solnica-Krezel 2005). In *Xenopus* embryos, overexpression of *Blimp1* downregulates the activity of Xbra and Myf5, proteins required for the formation of trunk mesoderm, and consequently causes axis truncations. Blimp1 also activates anterior mesoendodermal markers such as *goosecoid* and *cerberus* in animal cap explant assays. Similarly, zebrafish embryos treated with morpholino oligonucleotides show loss of anterior structures and display a foreshortened body axis (Wilm and Solnica-Krezel 2005). As stated above, germ cell differentiation in *Xenopus* and zebrafish is predetermined, and no defects in germ cell formation have so far been reported in these animals deficient in *Blimp1*. In mice, *Blimp1* is expressed in the analogous anterior visceral endoderm (AVE) and anterior definitive endoderm (ADE)/prechordal plate (Vincent et al. 2005). *Blimp1*-deficient embryos die at around E10.5, but early axis formation, anterior patterning, and neural crest formation proceed normally in these animals.

Blimp1 deficiency disrupts morphogenesis of the caudal branchial arches and leads to widespread blood leakage, tissue apoptosis and failure to correctly elaborate the labyrinthine layer of the placenta.

It is essential to identify the mechanism by which Blimp1 directs posterior proximal epiblast cells, which are otherwise destined to form somatic cells, toward PGC fate. Biochemical identification of a protein complex containing Blimp1 related to PGC specification would substantially increase our understanding of Blimp1 action. Since *Blimp1* expression continues in germ cells at least until E12.5 (Chang et al. 2002), Blimp1 may also be important for the maintenance of the germ cell phenotype, including preventing the reversion of germ cells into an explicitly totipotent state (see Sect. 11.4.1) or differentiation into somatic lineages.

11.4   Epigenetic Reprogramming in Early Germ Cells

Germline cells in mice undergo profound epigenetic reprogramming during the course of their development. This is most clearly demonstrated by the power of an oocyte to reprogram a somatic cell nucleus to complete full-term development (Wakayama et al. 1998, 2000). In normal germ cell development, parental imprints are erased in the PGCs in the genital ridges from E10.5 to E12.5 in both sexes (Hajkova et al. 2002). Coincidentally, genome-wide DNA demethylation, including in some repetitive elements occurs, and in females, the inactive X chromosome is reactivated (Kafri et al. 1992; Monk et al. 1987; Tam et al. 1994). The precise mechanisms regulating these events remain unresolved, however. We wish to understand the molecular basis of the germline reprogramming. Emerging data suggest that germ cells undergo extensive orderly reprogramming of epigenetic chromatin modifications soon after their fate is restricted (Seki et al. 2005).

11.4.1   Epigenetic Reprogramming
        in Migrating Primordial Germ Cells in Mice

We systematically analyzed the global states of key epigenetic modifications in migrating PGCs from E7.5 onward visualized by *Oct4 delta PE* EGFP, using whole mount immunohistochemistry (Seki et al. 2005).

Similar studies examining earlier germ cells from their emergence in the epiblast at the E6.5 to E7.5 EB stage are now underway using *Blimp1-mEGFP* mice.

Genome-Wide DNA Demethylation

As described earlier (see Sect. 11.3.1), the global level of cytosine methylation decreases through both an active and a passive mechanism until the blastocyst stage, and from this time on methylation increases due to de novo DNA methyltransferases Dnmt3b and Dnmt3a. The global cytosine methylation levels of PGCs at the base of the allantois at E8.0 or earlier is similar to those of their somatic neighbors, suggesting that PGCs and their somatic neighbors may originate from precursors with similar global DNA methylation levels. However, remarkably, PGCs migrating in the forming hindgut endoderm at E8.0 had much lower levels of cytosine methylation. This low level was maintained throughout their migration and was reduced further when the PGCs entered the gonads, reflecting the previous findings of genome-wide DNA demethylation and imprinting erasure in the genital ridges (Hajkova et al. 2002). These findings indicate that DNA demethylation in PGCs occurs in at least two distinct stages, the first at E8.0 in migrating PGCs and the second in gonadal PGCs.

We found that Dnmt3b and Dnmt3a are repressed in the founder and migrating PGCs. The maintenance methyltransferase Dnmt1 is transiently downregulated at the protein level in PGCs at around E8.0. Thus, DNA methyltransferase activity is at least transiently depleted from the PGCs' nucleus. It will be important to precisely determine the cell cycle state of the PGCs surrounding the genome-wide demethylation. Additionally, the sites of demethylation remain unknown, although the short interspersed nucleotide element (SINE) sequences would be the key candidates.

Genome-Wide Histone Tail Modifications

We examined the global distribution of H3K9me2, H3K27me3, H3K4me2, and H3-K9 acetylation of the migrating PGC genome.

H3-K9 Dimethylation   As with DNA methylation, founder PGCs at E7.5 showed genome-wide H3K9me2 similar to their somatic neighbors and precursor epiblast cells. This mark was distributed throughout the euchromatic regions with many punctate foci. Surprisingly, however, the global level of H3K9me2 decreased abruptly in PGCs around E8.0 (Fig. 4d–g), coincident with the genome-wide DNA demethylation seen in migrating PGCs. This suggests a mechanistic link between these two

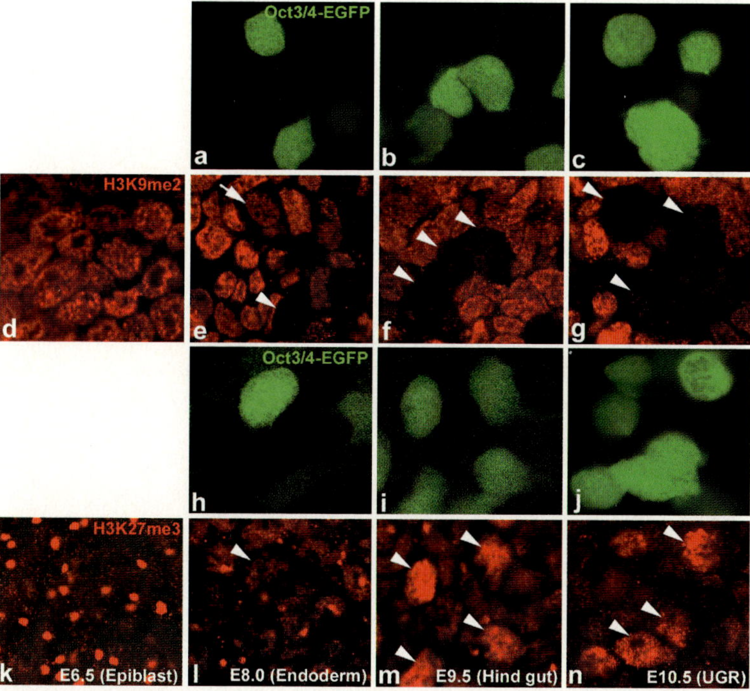

Fig. 4. a–n H3-K9 dimethylation and H3-K27 trimethylation states in developing PGCs compared to those in surrounding somatic cells. Oct4 delta PE GFP transgenic embryos at E6.5 (**d, k**), E8.0 (**a, e, h, l**), E9.5 (**b, f, i, m**), and E10.5 (**c, g, j, n**) stained with anti-H3K9me2 antibody (**d–g**) or anti-H3K27me3 antibody (**k–n**). GFP-positive cells (*green*, **a–c**, **h–i**) represent PGCs. In **e**, the *arrow* and *arrowhead* indicate a PGC with higher and lower H3K9me2, respectively. In **f**, **g**, **l–n**, *arrowheads* point to GFP-positive PGCs

events. The low level of H3K9me2 was maintained at least until E12.5 in PGCs.

We examined the distribution of HP1α, a factor that binds methylated H3K9, which mediates the formation of repressive chromatin (Bannister et al. 2001; Lachner et al. 2001). Consistent with the state of H3K9me2, the distribution of HP1α was similar in PGCs and their somatic neighbors until E8.0, but HP1α reactivity dramatically and specifically decreased in the euchromatic regions and the pericentric heterochromatin in PGCs at E9.5. Taken together, these findings demonstrate that migrating PGCs extensively reprogram their genome by erasing substantial levels of H3K9me2 around E8.0.

To the best of our knowledge, no other cell types exhibit such dramatic genome-wide reprogramming of histone modifications and DNA demethylation over a short period of time as seen in PGCs. The cell cycle state at which the reprogramming occurs is essential for a full mechanistic understanding of these events. As with DNA demethylation, it is important to determine the genomic regions in which H3K9me2 is erased. To accomplish this, however, powerful methods for determining the sequences of methylated DNA and histones must be developed, as the number of PGCs at this stage is limited. The functional significance of both DNA demethylation and changes in histone modifications remains unknown, and it is also unclear whether these events are mechanistically linked.

H3-K27 Trimethylation   H3-K27 methylation is a repressive modification mediated by polycomb group (PcG) proteins Ezh2/Eed complex (Erhardt et al. 2003; Plath et al. 2003; Silva et al. 2003). These two proteins are the most highly evolutionary conserved PcG proteins (Seydoux and Strome 1999). In mammals, these complexes are essential for early development, the initiation of X-inactivation and the derivation of ES cells as described earlier. H3-K27 methylation may be required for the plasticity seen in pluripotent cell types.

The global levels of H3K27me3 in founder PGCs are similar to that seen in their somatic neighbors until E8.5. At E8.75, however, these levels rise sharply in PGCs and are maintained at these levels until at least E12.5 (Fig. 4k–n). Current data suggest that PGCs convert repressive chromatin modifications from stable H3K9me2 to more plastic H3K27me3 to regain a chromatin state that is consistent with a totipotent state, with the concomitant expression of totipotency-associated transcription factors such as Oct4, Sox2, and nanog. Ezh2, Eed, and SUZ12 (Cao and Zhang 2004; Pasini et al. 2004), all essential for the observed H3K27me3, are expressed similarly in PGCs and their somatic neighbors at least until E8.25 (Saitou et al., unpublished data). It is essential to identify the mechanism regulating the timing and degree of genome-wide modifications seen in PGCs at E8.75. Continued expression of factors such as Blimp1 may prevent germline cells from both reverting to an explicitly totipotent phenotype seen in early blastomeres or ES cells and differentiating into somatic lineages.

H3-K4 Methylation and H3-K9 Acetylation   H3-K4 methylation and H3-K9 acetylation are coupled to transcriptionally permissive/active chromatin, and these two modifications mutually exclude H3-K9 methylation (Jenuwein and Allis 2001; Wang et al. 2001). We therefore examined the distribution of these two epigenetic marks. In E8.0 PGCs as well as in proximal epiblast cells, H3K4me2 was found throughout the euchromatin with many small foci, but it was largely absent from TOTO-3 positive pericentric heterochromatin and nucleoli. Nearby somatic cells exhibited similar staining patterns. However, levels of H3K4me2 appeared slightly higher in PGCs than in somatic cells. At E9.5, H3K4me2 levels were comparable between PGCs and somatic cells, but at E10.5 the levels of this modification were much higher in PGCs compared to surrounding gonadal somatic cells, although the signal distribution was indistinguishable between these cells. At E12.5, H3K4me2 levels again appeared similar in germ cells and somatic cells. Patterns of H3-K9 acetylation were very similar to the distribution of H3K4me2. H3-K9 acetylation levels appeared slightly higher in E8.0 PGCs, and staining of germ cells was much higher than surrounding somatic cells at E10.5. Thus, germ cells seem to adopt a transcriptionally permissive/active chromatin state similar to somatic cells, but at E10.5, the period when germ cells enter the genital ridge, they transiently establish a hyperactive/permissive chromatin pattern. However, these hyper-H3K4me2/H3-K9-acetylated states are not necessarily associated with

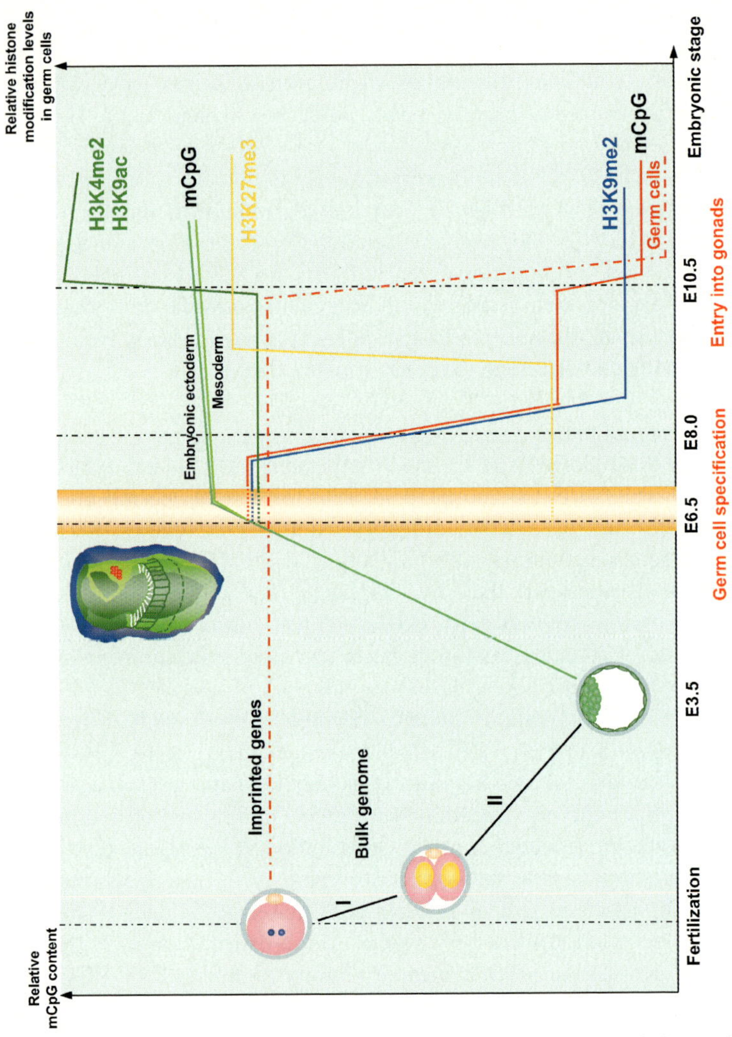

Fig. 5. Schematic representation of genome-wide reprogramming of chromatin modifications in germ cell development. Epigenetic reprogramming in PGCs is shown in combination with that during the developmental period up to germ cell specification. Only DNA methylation behavior is shown for the early stages. See text for details

hyperactive transcription; phosphorylation of serine-5 of the C-terminal domain (CTD) of RNA polymerase II, a modification associated with transcriptional elongation, was equivalent between gonadal PGCs and their somatic neighbors.

Collectively, we found that migrating PGCs erase two major repressive modifications, DNA cytosine methylation and H3K9me2, to a large extent from their euchromatic regions at around E8.0. There is also the elevation of H3K27me3 at E8.75, a more plastic, totipotency-associated modification. All of these changes are consistent with the genome of PGCs adopting a chromatin state associated with totipotency. In contrast, genome-wide modifications associated with transcriptionally permissive chromatin, H3K4me2 and H3-K9 acetylation, are acutely increased when PGCs enter the gonads at E10.5 (Fig. 5). These reprogramming events occur in a highly ordered manner and are likely associated with a genetic program of germline specification and development. A more precise description of the dynamics of these events is required, as well as the exploration of the functional significance of these events. The latter would include a search for mutations that affect this process with a concomitant defect in germ cell development.

11.4.2 Epigenetic States in Early Germ Cells in *Caenorhabditis elegans* and *Drosophila melanogaster*

It is important to note that the chromatin modifications that occur in early PGCs in mice described here are significantly different from those seen in *C. elegans* and *D. melanogaster*, two major model organisms with preformed germ cell determination. The increased complexity of epigenetic regulation seen in mice including the acquisition of parental imprinting and a substantially different mode of early development and germ cell specification may account for these differences.

In *C. elegans*, early germline precursors are transcriptionally inactive, with no phosphorylation of the C-terminal domains (CTD) at Ser-2 of RNA polymerase II (RNA pol.II) (Seydoux and Dunn 1997). However, these germline precursor blastomeres (P1–P3) exhibit transcriptionally permissive chromatin states with high levels of H3-K4 methylation across the genome, which is similar to the chromatin state in somatic blastomeres (Schaner et al. 2003). This is presumably because

P1–P3 precursors produce one somatic blastomere at each division that must begin somatic transcription soon after cleavage. Transcriptional quiescence in these germline precursors is instead maintained by the activity of a germline-specific Zn finger protein, PIE-1, which inhibits transcriptional elongation by competing with RNA pol.II CTD phosphorylation sites (Seydoux et al. 1996; Zhang et al. 2003). In contrast, the chromatin of the Z2 and Z3 cells that produce two germline cells after cell division exhibit a transcriptionally inactive state with high levels of H3-K9 methylation (Schaner et al. 2003). Thus, germline precursors in *C. elegans* shift their chromatin states from a transcriptionally permissive state to an inactive one at the P4 to Z2 and Z3 transition with concomitant degradation of PIE-1 protein.

The germline precursors in *D. melanogaster*, known as pole cells, are also transcriptionally inactive with no phosphorylation of CTD of RNA pol.II, as found in *C. elegans* (Schaner et al. 2003; Seydoux and Dunn 1997; Van Doren et al. 1998). However, in contrast to *C. elegans*, their chromatin adopts a transcriptionally inactive state at the outset of development with high levels of H3-K9 methylation. Pole cells produce only germline cells, and thus may be considered epigenetically similar to P4 or Z2 and Z3 cells in *C. elegans*. Recent studies have shown that the regulation of transcriptional silencing in these putative germ cells requires the *polar granule component* (*pgc*) gene (Deshpande et al. 2004; Martinho et al. 2004). Absence of *pgc* results in the loss of repression without affecting pole cell formation itself. The mutant pole cells display a precocious phosphorylation of ser2 of CTD. The mechanism of *pgc* is currently unknown, but it has been suggested that *pgc* might sequester critical components needed for PolII CTD phosphorylation for the transition from preinitiation complex to the elongation complex.

11.5 Conclusion and Perspectives

We have described here our recent progress on the understanding of germ cell specification in mice, together with epigenetic reprogramming associated with this event. Germline cells in mice originate from epiblast cells, which are also the precursors of all the somatic cells. All epiblast cells are destined to proceed toward somatic lineages, but the germline cells are recruited from this population through the activity of Blimp1, a potent transcriptional repressor with a modular structure exhibiting a variety of functions throughout development and in adults. The mechanism of Blimp1 function in germline precursors is not yet known, although we have evidence that *Blimp1*-deficient cells do not develop consistent expression of germ-cell markers including *stella*, *Sox2*, and *nanos3*, or the repression of *Hox* genes. However, as the effects of the loss of *Blimp1* appeared at a very early developmental stage (earlier than EB stage, presumably at the MS or LS stages), it has not been conclusively demonstrated that these defects were a direct consequence of Blimp1 deficiency. If Blimp1 suppresses a somatic program and leads to the specification of germ cell lineage, it will be an essential first step to precisely examine the nature of the somatic program in neighboring somatic cells.

In the course of our studies, we observed the restricted induction of *Blimp1* expression to a few cells in the most proximal layer of the epiblast on the posterior side (Figs. 2, 3). The mechanism governing this restricted expression is an area of active research. This is consistent with our findings that *Blimp1*-expressing cells at the ES stage show expression of posterior markers, including *nodal*, *Fgf8*, *cripto*, and *T*(*brachyury*) (Saitou et al., unpublished observation). It would therefore be conceivable that *Blimp1* induction may be repressed by the anteriorizing signals from the anterior visceral endoderm, which determines the anterior–posterior axis of the embryos (Rossant and Tam 2004). Our preliminary data also showed that *Blimp1* expression in proximal epiblast cells was completely abolished in the *Bmp4*-deficient embryos that completely lack PGCs. *Blimp1* expression was also severely affected in *Smad1*-null embryos, although some very weakly positive few cells were seen in some cases (Ohinata et al., unpublished observation). Examination of *Blimp1* expression on mutant backgrounds with defects in gastrulation and germ cell development will facilitate an understanding of the signaling cascades promoting germ cell specification. The determination of a possible enhancer element of *Blimp1* expression in the germline is also of great importance.

Our studies on the epigenetic states of migrating PGCs revealed that PGCs undergo extensive reprogramming of chromatin modifications in a highly ordered manner throughout the genome (Figs. 4, 5). This reprogramming begins with the elimination of H3K9me2 and DNA methylation at around E8.0, which is followed by increases in H3K27me3 at around E8.75. Since the erasure of both DNA and histone methylation occurs soon after the establishment of the germ cell lineage, it is reasonable to hypothesize that germ cell specification is a genetic process leading to the generation of the cell lineage capable of extensive epigenetic reprogramming toward a totipotent state, with the concomitant reacquisition of totipotency associated transcription factor networks consisting at least of *Oct4*, *Sox2*, and *nanog* (Saitou et al., unpublished observation). Continued expression of Blimp1 in PGCs may be critical to prevent reversion to an explicit totipotent phenotype or differentiation into some somatic lineage.

In this sense, a precise understanding of the mechanisms regulating germ cell specification will be essential for identifying the mechanisms governing epigenetic reprogramming. Towards this end, it is important to determine the cell cycle state of the germline cells from their emergence in the epiblast, since this knowledge is critical to identify if the erasure of DNA and histone methylation in PGCs at around E8.0 is an active or passive process. Our preliminary data suggest that the cell cycle states of early germ cells are tightly regulated (Seki et al., unpublished observation), which may act as a key for the observed reprogramming event. We found that all DNA methyltransferase activities are at least transiently eliminated from the germ cell nucleus during this critical period, and our preliminary data indicate that an essential component for euchromatic H3-K9 methylation is also repressed with similar kinetics (Seki et al., unpublished observation), both of which may be prerequisites for this event.

Our studies may lead to the development of a system to recapitulate the precise process of early germ cell development from ES cells (Geijsen et al. 2004; Hubner et al. 2003; Toyooka et al. 2003). These studies are extremely difficult because the relevant cell populations are very small and deeply embedded in different types of somatic cells throughout their development, making their purification and characterization difficult. The step-by-step understanding of germ cell formation should allow the in vitro generation of germ cells with appropriate epigenotype, which should enhance our understanding of the biochemical mechanism of epigenetic reprogramming. Such an understanding is profoundly important for the future of medicine in the post-genome era.

Acknowledgements. We thank all the members of our laboratory for discussion. We are particularly grateful to Professor Alexander Tarakovsky for providing us with the *Blimp1* mutant mice, Professor Michel Nussengzeig and Dr. Tetyana Obukhanych for the *Blimp1*-Cre mice. M.S. thanks Drs. Kirstie Lawson and Anne McLaren for discussion on the lineage restriction of PGCs. M.S. is supported in part by a Grant-in-Aid from the Ministry of Education, Culture, Sports, Science and Technology, Japan, and a PRESTO project grant from the Japan Science and Technology Agency.

References

Arney KL, Bao S, Bannister AJ, Kouzarides T, Surani MA (2002) Histone methylation defines epigenetic asymmetry in the mouse zygote. Int JDev Biol 46:317–320

Avilion AA, Nicolis SK, Pevny LH, Perez L, Vivian N, Lovell-Badge R (2003) Multipotent cell lineages in early mouse development depend on SOX2 function. Genes Dev 17:126–140

Bannister AJ, Zegerman P, Partridge JF, Miska EA, Thomas JO, Allshire RC, Kouzarides T (2001) Selective recognition of methylated lysine 9 on histone H3 by the HP1 chromo domain. Nature 410:120–124

Baxendale S, Davison C, Muxworthy C, Wolff C, Ingham PW, Roy S (2004) The B-cell maturation factor Blimp-1 specifies vertebrate slow-twitch muscle fiber identity in response to Hedgehog signaling. Nat Genet 36:88–93

Bird A (2002) DNA methylation patterns and epigenetic memory. Genes Dev 16:6–21

Blackwell TK (2004) Germ cells: finding programs of mass repression. Curr Biol 14, R229–R230

Brinster RL (2002) Germline stem cell transplantation and transgenesis. Science 296 2174–2176

Cao R, Zhang Y (2004) SUZ12 is required for both the histone methyltransferase activity and the silencing function of the EED-EZH2 complex. Mol Cell 15:57–67

Chang DH, Cattoretti G, Calame KL (2002) The dynamic expression pattern of B lymphocyte induced maturation protein-1 (Blimp-1) during mouse embryonic development. Mech Dev 117:305–309

Chang H, Matzuk MM (2001) Smad5 is required for mouse primordial germ cell development. Mech Dev 104:61–67

Chiquoine AD (1954) The identification, origin and migration of the primordial germ cells in the mouse embryo. Anat Rec 118:135–146

Chu GC, Dunn NR, Anderson DC, Oxburgh L, Robertson EJ (2004) Differential requirements for Smad4 in TGFbeta-dependent patterning of the early mouse embryo. Development 131:3501–3512

De Sousa Lopes SM, Roelen BA, Monteiro RM, Emmens R, Lin HY, Li E, Lawson KA, Mummery CL (2004) BMP signaling mediated by ALK2 in the visceral endoderm is necessary for the generation of primordial germ cells in the mouse embryo. Genes Dev 18:1838–1849

De Souza FS, Gawantka V, Gomez AP, Delius H, Ang SL, Niehrs C (1999) The zinc finger gene Xblimp1 controls anterior endomesodermal cell fate in Spemann's organizer. EMBO J 18 6062–6072

Dean W, Santos F, Stojkovic M, Zakhartchenko V, Walter J, Wolf E, Reik W (2001) Conservation of methylation reprogramming in mammalian development: aberrant reprogramming in cloned embryos. Proc Natl Acad Sci U S A 98:13734–3738

Deblandre GA, Marinx OP, Evans SS, Majjaj S, Leo O, Caput D, Huez GA, Wathelet MG (1995) Expression cloning of an interferon-inducible 17-kDa membrane protein implicated in the control of cell growth. J Biol Chem 270:23860–23866

Deshpande G, Calhoun G, Schedl P (2004) Overlapping mechanisms function to establish transcriptional quiescence in the embryonic Drosophila germline. Development 131:1247–1257

Dodge JE, Kang YK, Beppu H, Lei H, Li E (2004) Histone H3-K9 methyltransferase ESET is essential for early development. Mol Cell Biol 24:2478–2486

Eddy EM (1975) Germ plasm and the differentiation of the germ cell line. Int Rev Cytol 43:229–280

Erhardt S, Su IH, Schneider R, Barton S, Bannister AJ, Perez-Burgos L, Jenuwein T, Kouzarides T, Tarakhovsky A, Surani MA (2003) Consequences of the depletion of zygotic and embryonic enhancer of zeste 2 during preimplantation mouse development. Development 130:4235–4248

Evans MJ, Kaufman MH (1981) Establishment in culture of pluripotential cells from mouse embryos. Nature 292:154–156

Evans SS, Lee DB, Han T, Tomasi TB, Evans RL (1990) Monoclonal antibody to the interferon-inducible protein Leu-13 triggers aggregation and inhibits proliferation of leukemic B cells. Blood 76:2583–2593

Evans SS, Collea RP, Leasure JA, Lee DB (1993) IFN-alpha induces homotypic adhesion and Leu-13 expression in human B lymphoid cells. J Immunol 150:736–747

Extavour CG, Akam M (2003) Mechanisms of germ cell specification across the metazoans: epigenesis and preformation. Development 130:5869–5884

Gardner RL, Rossant J (1979) Investigation of the fate of 4–5 day post-coitum mouse inner cell mass cells by blastocyst injection. J Embryol Exp Morphol 52:141–152

Gardner RL, Lyon MF, Evans EP, Burtenshaw MD (1985) Clonal analysis of X-chromosome inactivation and the origin of the germ line in the mouse embryo. J Embryol Exp Morphol 88:349–363

Geijsen N, Horoschak M, Kim K, Gribnau J, Eggan K, Daley GQ (2004) Derivation of embryonic germ cells and male gametes from embryonic stem cells. Nature 427:148–154

Ginsburg M, Snow MH, McLaren A (1990) Primordial germ cells in the mouse embryo during gastrulation. Development 110:521–528

Gyory I, Wu J, Fejer G, Seto E, Wright KL (2004) PRDI-BF1 recruits the histone H3 methyltransferase G9a in transcriptional silencing. Nat Immunol 5:299–308

Hajkova P, Erhardt S, Lane N, Haaf T, El-Maarri O, Reik W, Walter J, Surani MA (2002) Epigenetic reprogramming in mouse primordial germ cells. Mech Dev 117:15–23

Hamatani T, Daikoku T, Wang H, Matsumoto H, Carter MG, Ko MS, Dey SK (2004) Global gene expression analysis identifies molecular pathways distinguishing blastocyst dormancy and activation. Proc Natl Acad Sci U S A 101:10326–10331

Hayashi K, Kobayashi T, Umino T, Goitsuka R, Matsui Y, Kitamura D (2002) SMAD1 signaling is critical for initial commitment of germ cell lineage from mouse epiblast. Mech Dev 118:99–109

Hernandez-Lagunas L, Choi IF, Kaji T, Simpson P, Hershey C, Zhou Y, Zon L, Mercola M, Artinger KB (2005) Zebrafish narrowminded disrupts the transcription factor prdm1 and is required for neural crest and sensory neuron specification. Dev Biol 278:347–357

Hubner K, Fuhrmann G, Christenson LK, Kehler J, Reinbold R, De La Fuente R, Wood J, Strauss JF 3rd, Boiani M, Scholer HR (2003) Derivation of oocytes from mouse embryonic stem cells. Science 300:1251–1256

Jenuwein T, Allis CD (2001) Translating the histone code. Science 293:1074–1080

Johnson AD, Crother B, White ME, Patient R, Bachvarova RF, Drum M, Masi T (2003) Regulative germ cell specification in axolotl embryos: a primitive trait conserved in the mammalian lineage. Philos Trans RSoc Lond B Biol Sci 358:1371–1379

Kafri T, Ariel M, Brandeis M, Shemer R, Urven L, McCarrey J, Cedar H, Razin A (1992) Developmental pattern of gene-specific DNA methylation in the mouse embryo and germ line. Genes Dev 6:705–714

Lachner M, Jenuwein T (2002) The many faces of histone lysine methylation. Curr Opin Cell Biol 14:286–298

Lachner M, O'Carroll D, Rea S, Mechtler K, Jenuwein T (2001) Methylation of histone H3 lysine 9 creates a binding site for HP1 proteins. Nature 410:116–120

Lachner M, O'Sullivan RJ, Jenuwein T (2003) An epigenetic road map for histone lysine methylation. J Cell Sci 116:2117–2124

Lane N, Dean W, Erhardt S, Hajkova P, Surani A, Walter J, Reik W (2003) Resistance of IAPs to methylation reprogramming may provide a mechanism for epigenetic inheritance in the mouse. Genesis 35:88–93

Lawson KA, Hage WJ (1994) Clonal analysis of the origin of primordial germ cells in the mouse. Ciba Found Symp 182:68–84

Lawson KA, Dunn NR, Roelen BA, Zeinstra LM, Davis AM, Wright CV, Korving JP, Hogan BL (1999) Bmp4 is required for the generation of primordial germ cells in the mouse embryo. Genes Dev 13:424–436

Leatherman JL, Jongens TA (2003) Transcriptional silencing and translational control: key features of early germline development. Bioessays 25:326–335

Lee J, Inoue K, Ono R, Ogonuki N, Kohda T, Kaneko-Ishino T, Ogura A, Ishino F (2002) Erasing genomic imprinting memory in mouse clone embryos produced from day 11.5 primordial germ cells. Development 129:1807–1817

Li E (2002) Chromatin modification and epigenetic reprogramming in mammalian development. Nat Rev Genet 3:662–673

Lin H (1997) The tao of stem cells in the germline. Annu Rev Genet 31:455–491

Martin GR (1981) Isolation of a pluripotent cell line from early mouse embryos cultured in medium conditioned by teratocarcinoma stem cells. Proc Natl Acad Sci U S A 78:7634–7638

Martinho RG, Kunwar PS, Casanova J, Lehmann R (2004) A noncoding RNA is required for the repression of RNApolII-dependent transcription in primordial germ cells. Curr Biol 14:159–165

Matsui Y, Zsebo K, Hogan BL (1992) Derivation of pluripotential embryonic stem cells from murine primordial germ cells in culture. Cell 70:841–847

McLaren A (1999) Signaling for germ cells. Genes Dev 13:373–376

Mitsui K, Tokuzawa Y, Itoh H, Segawa K, Murakami M, Takahashi K, Maruyama M, Maeda M, Yamanaka S (2003) The homeoprotein nanog is required for maintenance of pluripotency in mouse epiblast and ES cells. Cell 113:631–642

Monk M, Boubelik M, Lehnert S (1987) Temporal and regional changes in DNA methylation in the embryonic, extraembryonic and germ cell lineages during mouse embryo development. Development 99:371–382

Morgan HD, Santos F, Green K, Dean W, Reik W (2005) Epigenetic reprogramming in mammals. Hum Mol Genet 14:R47–R58

Nichols J, Zevnik B, Anastassiadis K, Niwa H, Klewe-Nebenius D, Chambers I, Scholer H, Smith A (1998) Formation of pluripotent stem cells in the mammalian embryo depends on the POU transcription factor Oct4. Cell 95:379–391

O'Carroll D, Erhardt S, Pagani M, Barton SC, Surani MA, Jenuwein T (2001) The polycomb-group gene Ezh2 is required for early mouse development. Mol Cell Biol 21:4330–4336

Ohinata Y, Payer B, O'Carroll D, Ancelin K, Ono Y, Sano M, Barton SC, Obukhanych T, Nussenzweig M, Tarakhovsky A et al (2005) Blimp1 is a critical determinant of the germ cell lineage in mice. Nature 436:207–213

Okano M, Bell DW, Haber DA, Li E (1999) DNA methyltransferases Dnmt3a and Dnmt3b are essential for de novo methylation and mammalian development. Cell 99:247–257

Olek A, Walter J (1997) The pre-implantation ontogeny of the H19 methylation imprint. Nat Genet 17:275–276

Pasini D, Bracken AP, Jensen MR, Lazzerini Denchi E, Helin K (2004) Suz12 is essential for mouse development and for EZH2 histone methyltransferase activity. EMBO J 23 4061–4071

Payer B, Saitou M, Barton SC, Thresher R, Dixon JP, Zahn D, Colledge WH, Carlton MB, Nakano T, Surani MA (2003) Stella is a maternal effect gene required for normal early development in mice. Curr Biol 13 2110–2117

Peters AH, O'Carroll D, Scherthan H, Mechtler K, Sauer S, Schofer C, Weipoltshammer K, Pagani M, Lachner M, Kohlmaier A et al. (2001) Loss of the Suv39h histone methyltransferases impairs mammalian heterochromatin and genome stability. Cell 107:323–337

Peters AH, Kubicek S, Mechtler K, O'Sullivan RJ, Derijck AA, Perez-Burgos L, Kohlmaier A, Opravil S, Tachibana M, Shinkai Y et al (2003) Partitioning and plasticity of repressive histone methylation states in mammalian chromatin. Mol Cell 12 1577–1589

Plath K, Fang J, Mlynarczyk-Evans SK, Cao R, Worringer KA, Wang H, de la Cruz CC, Otte AP, Panning B, Zhang Y (2003) Role of histone H3 lysine 27 methylation in X inactivation. Science 300:131–135

Ren B, Chee KJ, Kim TH, Maniatis T (1999) PRDI-BF1/Blimp-1 repression is mediated by corepressors of the Groucho family of proteins. Genes Dev 13:125–137

Rossant J, Tam PP (2004) Emerging asymmetry and embryonic patterning in early mouse development. Dev Cell 7:155–164

Rougier N, Bourc'his D, Gomes DM, Niveleau A, Plachot M, Paldi A, Viegas-Pequignot E (1998) Chromosome methylation patterns during mammalian preimplantation development. Genes Dev 12:2108–2113

Roy S, Ng T (2004) Blimp-1 specifies neural crest and sensory neuron progenitors in the zebrafish embryo. Curr Biol 14:1772–1777

Saitou M, Barton SC, Surani MA (2002) A molecular programme for the specification of germ cell fate in mice. Nature 418:293–300

Saitou M, Payer B, Lange UC, Erhardt S, Barton SC, Surani MA (2003) Specification of germ cell fate in mice. Philos Trans RSoc Lond B Biol Sci 358:363–370

Santos F, Hendrich B, Reik W, Dean W (2002) Dynamic reprogramming of DNA methylation in the early mouse embryo. Dev Biol 241:172–182

Santos F, Peters AH, Otte AP, Reik W, Dean W (2005) Dynamic chromatin modifications characterise the first cell cycle in mouse embryos. Dev Biol 280:225–236

Sato M, Kimura T, Kurokawa K, Fujita Y, Abe K, Masuhara M, Yasunaga T, Ryo A, Yamamoto M, Nakano T (2002) Identification of PGC7, a new gene expressed specifically in preimplantation embryos and germ cells. Mech Dev 113:91–94

Schaner CE, Deshpande G, Schedl PD, Kelly WG (2003) A conserved chromatin architecture marks and maintains the restricted germ cell lineage in worms and flies. Dev Cell 5:747–757

Sciammas R, Davis MM (2004) Modular nature of Blimp-1 in the regulation of gene expression during B cell maturation. J Immunol 172:5427–440

Seki Y, Hayashi K, Itoh K, Mizugaki M, Saitou M, Matsui Y (2005) Extensive and orderly reprogramming of genome-wide chromatin modifications associated with specification and early development of germ cells in mice. Dev Biol 278:440–458

Seydoux G, Dunn MA (1997) Transcriptionally repressed germ cells lack a subpopulation of phosphorylated RNA polymerase II in early embryos of *Caenorhabditis elegans* and *Drosophila melanogaster*. Development 124:191–201

Seydoux G, Strome S (1999) Launching the germline in *Caenorhabditis elegans*: regulation of gene expression in early germ cells. Development 126:3275–3283

Seydoux G, Mello CC, Pettitt J, Wood WB, Priess JR, Fire A (1996) Repression of gene expression in the embryonic germ lineage of *C. elegans*. Nature 382:713–716

Shaffer AL, Lin KI, Kuo TC, Yu X, Hurt EM, Rosenwald A, Giltnane JM, Yang L, Zhao H, Calame K et al. (2002) Blimp-1 orchestrates plasma cell differentiation by extinguishing the mature B cell gene expression program. Immunity 17:51–62

Shapiro-Shelef M, Lin KI, McHeyzer-Williams LJ, Liao J, McHeyzer-Williams MG, Calame K (2003) Blimp-1 is required for the formation of immunoglobulin secreting plasma cells and pre-plasma memory B cells. Immunity 19:607–620

Silva J, Mak W, Zvetkova I, Appanah R, Nesterova TB, Webster Z, Peters AH, Jenuwein T, Otte AP, Brockdorff N (2003) Establishment of histone h3 methylation on the inactive X chromosome requires transient recruitment of Eed-Enx1 polycomb group complexes. Dev Cell 4:481–495

Strumpf D, Mao CA, Yamanaka Y, Ralston A, Chawengsaksophak K, Beck F, Rossant J (2005) Cdx2 is required for correct cell fate specification and differentiation of trophectoderm in the mouse blastocyst. Development 132 2093–2102

Surani MA (2001) Reprogramming of genome function through epigenetic inheritance. Nature 414:122–128

Tachibana M, Sugimoto K, Nozaki M, Ueda J, Ohta T, Ohki M, Fukuda M, Takeda N, Niida H, Kato H et al. (2002) G9a histone methyltransferase plays a dominant role in euchromatic histone H3 lysine 9 methylation and is essential for early embryogenesis. Genes Dev 16:1779–1791

Tachibana M, Ueda J, Fukuda M, Takeda N, Ohta T, Iwanari H, Sakihama T, Kodama T, Hamakubo T, Shinkai Y (2005) Histone methyltransferases G9a and GLP form heteromeric complexes and are both crucial for methylation of euchromatin at H3-K9. Genes Dev 19:815–826

Tam PP, Zhou SX, Tan SS (1994) X-chromosome activity of the mouse primordial germ cells revealed by the expression of an X-linked lacZ transgene. Development 120:2925–2932

Toyooka Y, Tsunekawa N, Akasu R, Noce T (2003) Embryonic stem cells can form germ cells in vitro. Proc Natl Acad Sci U S A 100:11457–11462

Tremblay KD, Dunn NR, Robertson EJ (2001) Mouse embryos lacking Smad1 signals display defects in extra-embryonic tissues and germ cell formation. Development 128:3609–3621

Turner CA Jr, Mack DH, Davis MM (1994) Blimp-1, a novel zinc finger-containing protein that can drive the maturation of B lymphocytes into immunoglobulin-secreting cells. Cell 77:297–306

Van Doren M, Williamson AL, Lehmann R (1998) Regulation of zygotic gene expression in *Drosophila* primordial germ cells. Curr Biol 8:243–246

Vincent SD, Dunn NR, Sciammas R, Shapiro-Shalef M, Davis MM, Calame K, Bikoff EK, Robertson EJ (2005) The zinc finger transcriptional repressor Blimp1/Prdm1 is dispensable for early axis formation but is required for specification of primordial germ cells in the mouse. Development 132:1315–1325

Wakayama T, Perry AC, Zuccotti M, Johnson KR, Yanagimachi R (1998) Full-term development of mice from enucleated oocytes injected with cumulus cell nuclei. Nature 394:369–374

Wakayama T, Shinkai Y, Tamashiro KL, Niida H, Blanchard DC, Blanchard RJ, Ogura A, Tanemura K, Tachibana M, Perry AC et al. (2000) Cloning of mice to six generations. Nature 407:318–319

Wang H, Cao R, Xia L, Erdjument-Bromage H, Borchers C, Tempst P, Zhang Y (2001) Purification and functional characterization of a histone H3-lysine 4-specific methyltransferase. Mol Cell 8:1207–1217

Wilm TP, Solnica-Krezel L (2005) Essential roles of a zebrafish prdm1/blimp1 homolog in embryo patterning and organogenesis. Development 132:393–404

Wylie C (1999) Germ cells. Cell 96:165–174

Ying Y, Liu XM, Marble A, Lawson KA, Zhao GQ (2000) Requirement of Bmp8b for the generation of primordial germ cells in the mouse. Mol Endocrinol 14:1053–1063

Zhang F, Barboric M, Blackwell TK, Peterlin BM (2003) A model of repression: CTD analogs and PIE-1 inhibit transcriptional elongation by P-TEFb. Genes Dev 17:748–758

12 Transplantation of Germ Line Stem Cells for the Study and Manipulation of Spermatogenesis

I. Dobrinski

12.1   Introduction 176
12.2   Characterization of Germ Line Stem Cells 177
12.3   Germ Cell Culture 178
12.4   Germ Cell Transplantation to Study Phenotypes of Infertility . . . 179
12.5   Cross-Species Germ Cell Transplantation 180
12.6   Germ Cell Transplantation in Non-rodent Species 181
12.7   Transplantation of Testis Tissue 182
12.8   Applications for Transplantation of Germ Cells and Testis Tissue . 184
12.9   Conclusions 186
References 187

Abstract. Transplantation of male germ line stem cells from a fertile donor to the testis of an infertile recipient restores donor-derived spermatogenesis in the recipient testis and the resulting sperm pass the donor genotype to the offspring of the recipient. Germ cell transplantation has been an invaluable tool to elucidate the biology of male germ line stem cells and their niche in the testis, develop systems to isolate and culture spermatogonial stem cells, examine defects in spermatogenesis, correct male infertility and introduce genetic changes into the male germ line. Although most widely studied in rodents, germ cell transplantation has been applied to larger mammals, including primates. Recently, ectopic grafting of testis tissue from diverse donor species, including primates, into a mouse host has opened an additional possibility to study spermatogenesis and to produce fertile sperm from immature donors. Testis xenografts are ideally suitable to study toxicants or drugs with the potential to enhance or suppress male fertility without the necessity of performing experiments in the target species. Therefore, transplantation of germ cells or xenografting of testis tissue represent powerful approaches for the study, preservation, and manipulation of male fertility.

12.1 Introduction

Spermatogenesis is a continuous, highly organized process comprising sequential steps of cell proliferation and differentiation, resulting in the production of virtually unlimited numbers of spermatozoa throughout the life of the male (Russell et al. 1990). The foundation of this system is the spermatogonial stem cell, which has the potential for both self-renewal and production of differentiated daughter cells which will ultimately form spermatozoa (Huckins 1971; Clermont 1972; Meistrich and van Beek 1993). Among the stem cells in a male individual, the spermatogonial stem cell is unique in that it is the only cell in an adult body that divides and can contribute genes to subsequent generations. Because stem cells are ultimately defined by function, unequivocal identification depends on an assay to demonstrate the potential to reconstitute the appropriate body system.

This assay became available for spermatogonial stem cells when, in 1994, Dr. Ralph Brinster and colleagues at the University of Pennsylvania reported that transplantation of germ cells from fertile donor mice to the testes of infertile recipient mice results in donor-derived spermatogenesis and sperm production by the recipient animal (Brinster and Zimmermann 1994). The use of donor males carrying the bacterial β-galactosidase gene allowed for identification of donor-derived spermatogenesis in the recipient mouse testis and established the fact that donor haplotype is passed on to the offspring by recipient animals (Brinster and Avarbock 1994).

In the years following the initial report of the technique, several important steps were accomplished. In 1995, Jiang and Short applied the technique to germ cell transplantation between rats (subsequently also reported by Ogawa et al. (1999b) and Zhang et al. (2003), and in 1996, Brinster's group showed that mouse spermatogonial stem cells can be cryopreserved for prolonged periods of time before transplantation and still establish spermatogenesis in the recipient testis (Avarbock et al. 1996), thereby providing a means to virtually immortalize the genetic potential of a male individual. In the following year, a detailed technical analysis of the technique was published (Ogawa et al. 1997) and in 1999, we established an image analysis approach that now allows quantification of colonization of recipient testes by donor stem cells (Dobrinski et al. 1999b).

12.2 Characterization of Germ Line Stem Cells

With the fundamental aspects of germ cell transplantation in place, it became possible to study the stem cell niche in the testis and to characterize putative spermatogonial stem cells. The pattern and kinetics of colonization after transplantation were described in detail (Parreira et al. 1998; Nagano et al. 1999; Ventela et al. 2002), and cross-species transplantation showed that cell cycle during spermatogenesis is controlled by the germ cell, not the Sertoli cell (Franca et al. 1998). Sperm arising from transplanted donor germ cells were capable of fertilization in vivo and in vitro (Brinster and Avarbock 1994; Goossens et al. 2003; Honaramooz et al. 2003b).

It has been estimated that there are only about 2×10^4 stem cells in 10^8 cells of a mouse testis (Meistrich and van Beek 1993; Tegelenbosch and de Rooij 1993). In order to study the biology of these stem cells, it was therefore desirable to obtain populations of cells enriched in spermatogonial stem cells from testis cell preparations. Work in the mouse focused on characterization of and enrichment for putative stem cells. Initially it was shown that selection of germ cells for expression of α_6-and β_1-integrin the absence of c-kit receptor, as well as collection of cells from experimentally induced cryptorchid testes resulted in a significant enrichment for spermatogonial stem cells (Shinohara et al. 1999, 2000a, b). A subsequent study further characterized the antigenic phenotype of putative spermatogonial stem cells (Kubota et al. 2003); however, there were conflicting reports whether side-population cells that actively exclude Hoechst dye represent a cell population enriched in germ line stem cells as reported for other stem cell systems (Kubota et al. 2003; Lassalle et al. 2004). More recently, expression of Thy-1, CD9, or Egr3 surface proteins, also expressed on embryonic or bone marrow stem cells, were utilized as markers for enrichment of mouse germ line stem cell populations (Kubota et al. 2003, 2004; Kanatsu-Shinohara et al. 2004a; Hamra et al. 2004).

The success of germ cell transplantation requires the availability of a stem cell niche in the recipient testis. Analysis of different recipient models by choosing animal age and hormonal manipulation of the recipient testicular environment provided insight into characteristics of the stem cell niche in the testis.

Using young mice and treating recipients with GnRH agonists to suppress high intratesticular testosterone levels improved donor cell colonization (Shinohara et al. 2001; Ogawa et al. 1998, 1999b; Dobrinski et al. 2001). Interestingly, germ cell transplantation in rodents and perhaps also cattle requires that donor and recipients are closely related or that recipient animals are immunosuppressed (Kanatsu-Shinohara et al. 2003a; Zhang et al. 2003; Izadyar et al. 2003), whereas germ cell transplantation in pigs and goats was also successful between unrelated individuals (Honaramooz et al. 2002a, 2003a, b). Serial transplantation experiments have been employed to study stem cell proliferation and homing efficiency after transplantation into recipient mouse testes (Ogawa et al. 2003; Nagano 2003; Kanatsu-Shinohara et al. 2003c). More recently, a morphological characterization of stem cells and the development of the stem cell niche have also been presented for the rat (Orwig et al. 2002b; Ryu et al. 2003, 2004).

12.3   Germ Cell Culture

Nagano et al. (1998) showed first that stem cells could be maintained in culture for a long period of time. Improving culture conditions for male germ line stem cells still is under intense study, as evidenced by recent reports of improved culture systems for mouse germ cells (Kanatsu-Shinohara et al. 2003b; Jeong et al. 2003). Co-culture with embryonic fibroblast or bone marrow stromal cells, but not Sertoli cell lines, and addition of several growth factors known to be beneficial for culture of other stem cell types or primordial germ cells, such as glial cell-line-derived neurotrophic factor (GDNF), leukemia inhibitory factor (LIF), epidermal growth factor (EGF), and basic fibroblast growth factor (bFGF), successfully maintained mouse germline stem cells in culture for varying periods of time (Nagano et al. 1998, 2003; Kanatsu-Shinohara et al. 2003b). Recently, two groups reported efficient long-term culture system for mouse spermatogonial stem cells (Kubota et al. 2004; Kanatsu-Shinohara et al. 2005).

Transplantation experiments demonstrated the developmental potential of mouse primordial germ cells to initiate spermatogenesis when transplanted into a postnatal testis (Chuma et al. 2005) and subsequently it was reported that pluripotent stem cells could be isolated from cultures of neonatal mouse testis cells (Kanatsu-Shinohara et al. 2004b). It appears that multipotent cells persist in the neonatal mouse testis that will preferentially proliferate under specific culture conditions as was previously reported for multipotent cells isolated from porcine fetal fibroblast cultures (Dyce et al. 2004; Kues et al. 2005).

The majority of the research to date was conducted with primary cultures of putative male germ line stem cells. While progress in this area has been significant, availability of immortalized cell lines would provide tremendous potential for the study and manipulation of male germ cells in vitro. To date, there are reports of immortalized germ cell lines from rat and mouse (van Pelt et al. 2002; Feng et al. 2002; Hofmann et al. 2005).

12.4 Germ Cell Transplantation to Study Phenotypes of Infertility

Beginning in 2000, germ cell transplantation in rodents was increasingly employed to investigate whether a phenotype of infertility originated from a defect in Sertoli cells or in germ cells. Initially, we demonstrated that transplantation of germ cells from infertile Steel mutant mice to infertile W/W^v mutant mice could restore fertility (Ogawa et al. 2000). Subsequently, germ cell transplantation was used to characterize the role of c-kit and SCF in regulation of germ cell proliferation

(Ohta et al. 2000), to show that the defect associated with the juvenile spermatogonial depletion (*jsd*) mutation was inherent to the germ cells (Boettger-Tong et al. 2000; Ohta et al. 2001), that germ cells do not require estrogen receptors (Mahato et al. 2000) or androgen receptors (Johnston et al. 2001) for development, and that germ cell differentiation is regulated by GDNF (Creemers et al. 2002; Yomogida et al. 2003) and CREM function (Wistuba et al. 2002). Transplantation of wild-type germ cells into the testes of Dazl-null mice established that the somatic compartment of the Dazl-null testes remains functional (Rilianawati et al. 2003). More recently, plzf has been shown by reciprocal transplantation experiments using mutant and wild-type strains to be essential for stem cell self-renewal in the mouse (Buaas et al. 2004; Costoya et al. 2004). In the rat, germ cell transplantation experiments elucidated the defect underlying the as-mutation (Noguchi et al. 2002). This list is continuously growing as new rodent models with male infertility phenotypes become available.

12.5   Cross-Species Germ Cell Transplantation

In 1996, production of rat sperm in mouse testes was achieved following cross-species (xenogeneic) spermatogonial transplantation from rats to mice (Clouthier et al. 1996) and was subsequently successful from mice to rats (Ogawa et al. 1999b; Zhang et al. 2003). For its obvious practical potential, cross-species germ cell transplantation was explored further (Table 1). Hamster spermatogenesis occurred successfully in the mouse host (Ogawa et al. 1999a); however, with increasing phylogenetic distance between donor and recipient species, complete spermatogenesis could no longer be achieved in the mouse testis. Transplantation of germ cells from non-rodent donors ranging from rabbits and dogs, to pigs and bulls, and ultimately non-human primates and humans, resulted in colonization of the mouse testis, but spermatogenesis became arrested at the stage of spermatogonial expansion (Dobrinski et al. 1999a, 2000; Nagano et al. 2001b, 2002a). It appears that the initial steps of germ cell recognition by the Sertoli cells, localization to the basement membrane, and initiation of spermatogonial proliferation are conserved between evolutionary divergent species. However, it was hypothesized

Table 1. Cross-species transplantation of germ cells

Donor species	Recipient species	Donor-derived spermatogenesis	Reference
Rat	Mouse	Complete	Clouthier et al. 1996; Russell et al. 1996
Hamster	Mouse	Complete	Ogawa et al. 1999b
Rabbit, dog	Mouse	Spermatogonial proliferation but no further differentiation	Dobrinski et al. 1999a
Bull	Mouse	Spermatogonial proliferation but no further differentiation	Dobrinski et al. 2000; Izadyar et al. 2002
Pig, horse	Mouse	Spermatogonial proliferation but no further differentiation	Dobrinski et al. 2000
Baboon	Mouse	Spermatogonial proliferation but no further differentiation	Nagano et al. 2001b
Human	Mouse	No colonization; spermatogonial proliferation but no further differentiation	Reis et al. 2000; Nagano et al. 2002b
Mouse	Rat	Complete	Ogawa et al. 1999b
Mouse	Pig	No colonization	Honaramooz et al. 2002a

that with increasing phylogenetic distance between donor and recipient species, the recipient testicular environment (comprised of Sertoli cells and paracrine factors) becomes unable to support spermatogenic differentiation and meiosis. This incompatibility of donor germ cells and recipient testicular environment was overcome by co-transplantation of germ cells and Sertoli cells (Shinohara et al. 2003) or by testis tissue transplantation (Honaramooz et al. 2002b). Although xenogeneic spermatogonial transplantation did not have the envisioned immediate practical application, it nonetheless provides a bioassay for stem cell potential of germ cells isolated from other species (Dobrinski et al. 1999a, 2000; Izadyar et al. 2002) as well as an intriguing tool to study cellular and noncellular requirements of spermatogenesis in diverse species.

12.6   Germ Cell Transplantation in Non-rodent Species

While the majority of the research is conducted in rodent models, germ cell transplantation is now also applied to non-rodent species such as pigs, goats, cattle, and primates (Honaramooz et al. 2002a, 2003a, b;

Izadyar et al. 2003; Schlatt et al. 2002a). Application of germ cell transplantation technology to non-rodent species has been difficult. Due to differences in testicular anatomy and physiology, germ cells cannot be delivered by the same technique as in rodents. Instead, by combining ultrasound-guided cannulation of the centrally located rete testis with delivery of germ cells by gravity flow, we succeeded in transplanting donor cells from transgenic donor goats into the testes of immunocompetent, prepubertal recipient animals (Honaramooz et al. 2003a). Once these goats became sexually mature, they produced sperm carrying the donor haplotype and transmitted the donor genetic makeup to the offspring. This provided proof-of-principle that germ cell transplantation results in donor-derived sperm production and fertility in a non-rodent species. Importantly, donors and recipients were unrelated and immunocompetent. In most studies in rodents, syngeneic animals are used as donors and recipients or recipient animals are chosen from immunocompromised strains. Recent studies demonstrated that allogeneic spermatogenesis is possible in rodents, however, only when recipient animals received immunosuppressive treatment (Zhang et al. 2003; Kanatsu-Shinohara et al. 2003a). The testis is considered to be an immune privileged site, but it is unclear why transplantation between unrelated, immunocompetent animals is possible in domestic animal species but not in rodents. Nonetheless, this makes the technique infinitely more applicable in non-rodent species.

12.7 Transplantation of Testis Tissue

As outlined above, transplantation of isolated male germ cells is a very powerful approach for the study of spermatogenesis as well as for preservation and manipulation of the male germ line. However, the technique cannot be easily adapted between diverse species. Xenogeneic germ cell transplantation did not result in complete sperm production in species other than rodents, probably due to an incompatibility between donor germ cells and recipient Sertoli cells and testicular environment. In contrast, co-transplantation of the donor germ cells with their surrounding testicular tissue into a mouse host preserves the testicular integrity and still allows experimentation in a small rodent. We therefore developed ectopic xenografting of testicular tissue under the back skin of immun-odeficient mice as an alternate approach for the maintenance and prop-agation of male germ cells that can be more readily applied to different mammalian species (Honaramooz et al. 2002b). Testis tissue xenograft-ing maintains structural integrity of the testicular tissue and provides the accessibility essential for the study and manipulation of testis function as well as for male germ line preservation. Through this approach, we showed that xenografting of testis tissue from newborn pigs and goats resulted in production of normal, functional sperm in a mouse host (Fig. 1, Honaramooz et al. 2002b). This was the first report of complete, functional xenogeneic spermatogenesis in species other than rodents and it is also the first time that sperm could be obtained from neonatal donors. Subsequently, ectopic xenografting of testicular tissue was also reported from hamsters or marmoset monkeys to mice (Schlatt et al. 2002b). Similar to isolated germ cells, testicular tissue can be stored frozen prior to grafting and with retention of its developmental potential (Honaramooz et al. 2002b; Schlatt et al. 2002b; Shinohara et al. 2002). Sperm recovered from allografts (mouse) and xenografts (pig, goat) sup-ported pronuclear formation when injected into mouse oocytes. Mouse sperm recovered from allografts resulted in normal, fertile progeny when injected into mouse oocytes followed by embryo transfer (Schlatt et al. 2003). The onset of spermatogenesis in xenografted pig testis tissue occurred slightly earlier than in the donor species (Honaramooz et al. 2002a) and we subsequently demonstrated that testicular maturation and sperm production in rhesus macaques can be significantly accelerated by xenografting of testis tissue (Honaramooz et al. 2004) and results in production of sperm that support embryo development after injection into rhesus monkey oocytes.

The accessibility of the tissue in the mouse host makes it possible to manipulate spermatogenesis and steroidogenesis in a controlled manner that is not feasible in the donor animal and certainly not in humans. This in turn will allow analysis of the effects of toxicants and potential male contraceptives on testis function in an in vivo culture system without extensive experimentation in the target species.

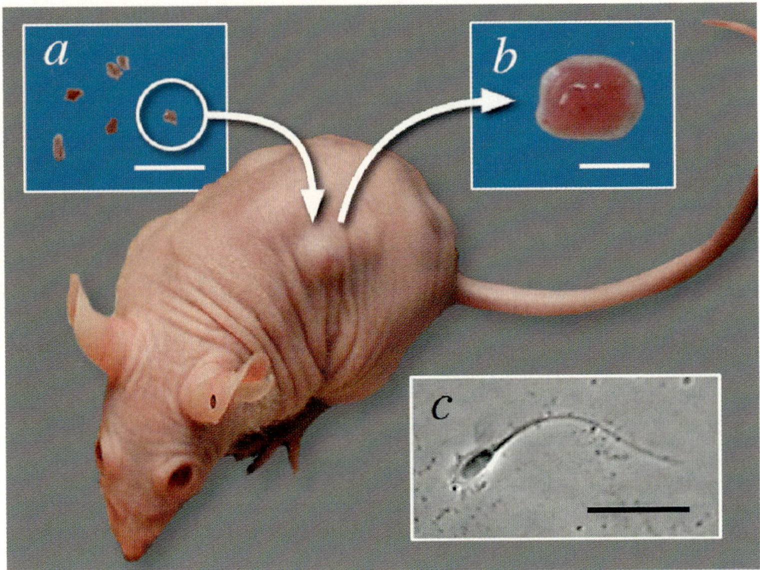

Fig. 1. a–c Ectopic xenografting of testis tissue from newborn piglets to nude mice. **a**: Small pieces of pig testis tissue (0.5–1 mm in diameter) were grafted under the back skin of nude mice. **b**: The grafts had expanded to 4–8 mm in diameter by 10 weeks after grafting. *c*: A porcine sperm recovered from a xenograft 27 weeks after grafting. Bars, 5 mm (**a, b**) or 2 µm (**c**). (Reprinted from Honaramooz et al. 2002b)

12.8 Applications for Transplantation of Germ Cells and Testis Tissue

Germ cell transplantation could serve to restore male fertility after an insult to the testis (Fig. 2). Specifically, one could preserve germ cells prior to irradiation or chemotherapy treatment for cancer as these treatments often lead to temporary or permanent destruction of spermatogenesis. Re-introduction of autologous germ cells could then restore fertility in the patient once the previous illness has been overcome. This approach was demonstrated in principle in the monkey (Schlatt et al. 2002a) and its application in humans has been discussed (Radford et al. 1999, 2003;

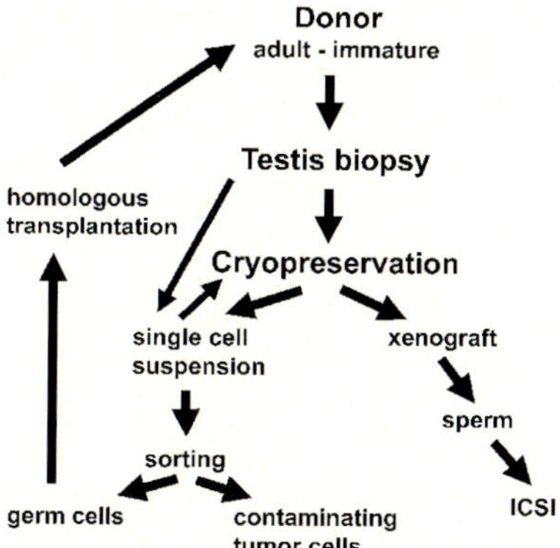

Fig. 2. Illustration of potential application of germ cell transplantation and testis tissue xenografting to the preservation or restoration of donor fertility. Testis tissue collected from an individual donor can be cryopreserved and stored. For homologous germ cell transplantation, a single cell suspension is prepared, either before or after cryostorage. Potential contaminating tumor cells can be removed by cell sorting and the germ cells can be transplanted back to the donor. Alternatively, testis tissue can be used for xenografting and the resulting sperm for fertilization by ICSI

Brook et al. 2001), but so far no definitive data have been reported. Re-introduction of donor cells collected before treatment carries the risk that cancerous cells could also be re-introduced into the patient as demonstrated in leukemic rats (Jahnukainen et al. 2001). Cell-sorting technology could potentially be employed to overcome this risk (Fujita et al. 2005). Germ cell transplantation has an advantage over the cryopreservation of sperm prior to treatment, in that it could be applied to prepubertal males where sperm cannot be obtained or to adult males rendered azoospermic or teratozoospermic by disease. The technique is also of great interest in domestic or endangered animals for its potential to preserve genetic material from immature males that are lost before they reach puberty.

Another important potential application of the technique is transplantation of transfected germ cells as an alternate means to generate transgenic animal models through the manipulation of the male germ line. Although the lack of pure starting populations of germ line stem cells and optimized culture systems together with the low proliferating activity of stem cells have made this a difficult task, some success has been reported in generating transgenic mice and rats by retroviral or lentiviral transduction of germ cells prior to transplantation (Nagano et al. 2001a, 2002b; Hamra et al. 2002; Orwig et al. 2002a). This approach has tremendous potential mainly in species such as the rat and domestic animals where embryonic stem cell technology is not available.

Ectopic testis tissue grafting is a new option for male germ line preservation. As it provides a potentially inexhaustible source of male gametes even from immature gonads, grafting of fresh or preserved testis tissue offers an invaluable tool for the conservation of fertility by allowing sperm production from immature males. Unlike autologous transplantation of isolated germ cells to restore fertility in a patient following cancer therapy, xenologous grafting and use of the resultant sperm for assisted fertilization would eliminate the potential risk of tumor cell transmission (Fig. 2). However, safety and ethical issues associated with the use of a xenogeneic system would have to be addressed before application to human patients is considered.

12.9   Conclusions

Germ cell transplantation, and more recently testis tissue xenografting, have provided a multitude of insights into male germ line stem cell biology and testis function. Transplantation systems can aid in elucidating factors controlling stem cell proliferation and differentiation and aspects of the stem cell niche in the testis. These approaches are also extensively used to study basic biological questions underlying the cause of male infertility. Practical applications are now developed to introduce genetic modifications into the germline of animals, and to preserve fertility in rare and endangered animals and oncological patients undergoing potentially sterilizing treatments for cancer therapy. The introduction of germ cell transplantation, especially in combination with recent progress in enrichment and culture of spermatogonial stem cells, has opened up seemingly endless possibilities to study and manipulate male germ line stem cells in mammalian species. It can be expected that germ cell transplantation will continue to significantly enhance our understanding of testis function and our ability to control and preserve male fertility.

Acknowledgements. Work from the author's laboratory presented here was supported by 5R01RR017359-04 1 R21 RR017419-01 (NCRR/NIH) 1R41 HD044780-01 (NICHD/NIH), and USDA/CSREES/NRICGP (2003-35203-13486).

References

Avarbock MR, Brinster CJ, Brinster RL (1996) Reconstitution of spermatogenesis from frozen spermatogonial stem cells. Nature Med 2:693–696

Boettger-Tong HL, Johnston DS, Russell LD, Griswold MD, Bishop CE (2000) Juvenile spermatogonial depletion (jsd) mutant seminiferous tubules are capable of supporting transplanted spermatogenesis Biol Reprod 63:1185–1191

Brinster RL, Avarbock MR (1994) Germline transmission of donor haplotype following spermatogonial transplantation. Proc Natl Acad Sci U S A 91:11303–11307

Brinster RL, Zimmermann JW (1994) Spermatogenesis following male germ-cell transplantation. Proc Natl Acad Sci U S A 91:11298–11302

Brook PF, Radford JA, Shalet SM, Joyce AD, Gosden RG (2001) Isolation of germ cells from human testicular tissue for low temperature storage and autotransplantation. Fertil Steril 75:269–274

Buaas FW, Kirsh AL, Sharma M, McLean DJ, Morris JL, Griswold MD, de Rooij DG, Braun RE (2004) Plzf is required in adult male germ cells for stem cell self-renewal. Nat Genet 36:647–652

Chuma S, Kanatsu-Shinohara M, Inoue K, Ogonuki N, Miki H, Toyokuni S, Hosokawa M, Nakatsuji N, Ogura A, Shinohara T (2005) Spermatogenesis from epiblast and primordial germ cells following transplantation into postnatal mouse testis. Development: 132:117–122

Clermont Y (1972) Kinetics of spermatogenesis in mammals: seminiferous epithelium cycle and spermatogonial renewal. Physiol Rev 52:198–236

Clouthier DE, Avarbock MR, Maika SD, Hammer RE, Brinster RL (1996) Rat spermatogenesis in mouse testis. Nature 381:418–421

Costoya JA, Hobbs RM, Barna M, Cattoretti G, Manova K, Sukhwani M, Orwig KE, Wolgemuth DJ, Pandolfi PP (2004) Essential role of Plzf in maintenance of spermatogonial stem cells. Nat Genet 36:653–659

Creemers LB, Meng X, den Ouden K, van Pelt MM, Izadyar F, Santoro M, de Rooij DG, Sariola H (2002) Transplantation of germ cells from glial cell line-derived neurotrophic factor-overexpressing mice to host testes depleted of endogenous spermatogenesis by fractionated irradiation. Biol Reprod 66:1579–1584

Dobrinski I, Avarbock MR, Brinster RL (1999a) Transplantation of germ cells from rabbits and dogs into mouse testes. Biol Reprod 61:1331–1339

Dobrinski I, Ogawa T, Avarbock MR, Brinster RL (1999b) Computer-assisted image analysis to assess colonization of recipient seminiferous tubules by spermatogonial stem cells from transgenic donor mice. Mol Reprod Dev 53:142–148

Dobrinski I, Avarbock MR, Brinster RL (2000) Germ cell transplantation from large domestic animals into mouse testes. Mol Reprod Develop 57:270–279

Dobrinski I, Ogawa T, Avarbock MR, Brinster RL (2001) Effect of the GnRH-agonist leuprolide on colonization of recipient testes by donor spermatogonial stem cells after transplantation in mice. Tissue Cell 33:200–207

Dyce PW, Zhu H, Craig J, Li J (2004) Stem cells with multilineage potential derived from porcine skin. Biochem Biophys Res Com 316:651–658

Feng L-X, Chen Y, Dettin L, Reijo Pera RA, Herr JC, Goldberg E, Dym M (2002) Generation and in vitro differentiation of a spermatogonial cell line. Science 297:392–395

Fujita K, Ohta H, Tsujimura A, Takao T, Miyagawa Y, Takada S, Matsumiya K, Wakayama T, Okuyama A (2005) Transplantation of spermatogonial stem cells isolated from leukemic mice restores fertility without inducing leukemia. J Clin Invest 115 :1855–1861

Franca LR, Ogawa T, Avarbock MR, Brinster RL, Russell LD (1998) Germ cell genotype controls cell cycle during spermatogenesis in the rat. Biol Reprod 59:1371–1377

Goossens E, Frederickx V, De Block G, Van Steirteghem A, Tournaye H (2003) Reproductive capacity of sperm obtained after germ cell transplantation in a mouse model. Hum Reprod 18:1874–1880

Hamra FK, Gatlin J, Chapman KM, Grellhesl DM, Garcia JV, Hammer RE, Garbers DL (2002) Production of transgenic rats by lentiviral transduction of male germ-line stem cells. Proc Natl Acad Sci U S A 99:14931–14936

Hamra FK, Schultz N, Chapman KM, Grellhesl DM, Cronkhite JT, Hammer RE, Garbers DL (2004) Defining the spermatogonial stem cell. Dev Biol 269:393–410

Hofmann M-C, Braydich-Stolle L, Dettin L, Johnson E, Dym M (2005) Immortalization of mouse germ line stem cells. Stem Cells 23:200–210

Honaramooz A, Megee SO, Dobrinski I (2002a) Germ cell transplantation in pigs. Biol Reprod 60:21–28

Honaramooz A, Snedaker A, Boiani M, Scholer HR, Dobrinski I, Schlatt S (2002b) Sperm from neonatal mammalian testes grafted in mice. Nature 418:778–781

Honaramooz A, Behboodi E, Blash, S, Megee SO, Dobrinski I (2003a) Germ cell transplantation in goats. Mol Reprod Dev 64:422–428

Honaramooz A, Behboodi E, Megee SO, Overton SA, Galantino-Homer HL, Echelard Y, Dobrinski I (2003b) Fertility and germline transmission of donor haplotype following germ cell transplantation in immunocompetent goats. Biol Reprod 69:1260–1264

Honaramooz A, Li M-W, Penedo MCT, Meyers SA, Dobrinski I (2004) Accelerated maturation of primate testis by xenografting into mice. Biol Reprod 70:1500–1503

Huckins C (1971) The spermatogonial stem cell population in adult rats. I. Their morphology, proliferation, and maturation. Anat Rec 160:533–558

Izadyar F, den Ouden K, Stout TAE, Stout J, Coret J, Lankveld DPK, Spoormakers TJP, Colenbrander B, Oldenbroek JK, Van der Ploeg KD, Woelders H, Kal HB, de Rooij DG (2003) Autologous and homologous transplantation of bovine spermatogonial stem cells. Reproduction 126:765–774

Izadyar F, Matthijs-Rijsenbilt JJ, den Ouden K, Creemers LB, Woelders H, de Rooij D (2002). Development of a cryopreservation protocol for type A spermatogonia. J Androl 23:537–545

Jahnukainen K, Hou M, Petersen C, Setchell B, Soder O (2001) Intratesticular transplantation of testicular cells from leukemic rats causes transmission of leukemia. Cancer Res 61:706–710

Jeong D, McLean DJ, Griswold MD (2003) Long-term culture and transplantation of murine testicular germ cells. J Androl 24:661–669

Jiang FX, Short RV (1995) Male germ cell transplantation in rats: apparent synchronization of spermatogenesis between host and donor seminiferous epithelia. Int J Androl 18:326–330

Johnston DS, Russell LD, Friel PJ, Griswold MD (2001) Murine germ cells do not require functional androgen receptors to complete spermatogenesis following spermatogonial stem cell transplantation. Endocrinology 142:2405–2408

Kanatsu-Shinohara M, Ogonuki N, Inoue K, Ogura A, Toyokuni S, Honjo T, Shinohara T (2003a) Allogeneic offspring produced by male germ line stem cell transplantation into infertile mouse testis. Biol Reprod 68:167–173

Kanatsu-Shinohara M, Ogonuki N, Inoue K, Miki H, Ogura A, Toyokuni S, Shinohara T (2003b) Long-term proliferation in culture and germline transmission of mouse male germline stem cells. Biol Reprod 69:612–616

Kanatsu-Shinohara M, Toyokuni S, Morimoto T, Matsui S, Honjo T, Shinohara T (2003c) Functional assessment of self-renewal activity of male germline stem cells following cytotoxic damage and serial transplantation. Biol Reprod 68:1801–1807

Kanatsu-Shinohara M, Toyokuni S, Shinohara T (2004a) CD9 is a surface marker on mouse and rat male germline stem cells. Biol Reprod 70:70–75

Kanatsu-Shinohara M, Inoue K, Lee J, Yoshimoto M, Ogonuki N, Miki H, Baba S, Kato T, Kazuki Y, Toyokuni S, Toyoshima M, Niwa O, Oshimura M, Heike T, Nakahata T, Ishino F, Ogura A, Shinohara T (2004b) Generation of pluripotent stem cells from neonatal mouse testis. Cell 119:1001–1012

Kanatsu-Shinohara M, Miki H, Inoue K, Ogonuki N, Toyokuni S, Ogura A, Shinohara T (2005) Long-term culture of mouse male germline stem cells under serum- or feeder-free conditions. Biol Reprod 72:985–991

Kubota H, Avarbock MR, Brinster RL (2003) Spermatogonial stem cells share some, but not all, phenotypic and functional characteristics with other stem cells. Proc Natl Acad Sci U S A 100:6487–6492

Kubota H, Avarbock MR, Brinster RL (2004) Culture conditions and single growth factors affect fate determination of mouse spermatogonial stem cells. Biol Reprod 71:722–731

Kues WA, Petersen B, Mysegades W, Carnwath JW, Niemann H (2005) Isolation of murine and porcine fetal stem cells from somatic tissue. Biol Reprod 72:1020–1028

Lassalle B, Bastos H, Louis JP, Testart J, Dutrillaux B, Fouchet P, Allemand I (2004) 'Side population' cells in adult mouse testis express Bcrp1 gene and are enriched in spermatogonia and germinal stem cells. Development 131:479–487

Mahato D, Goulding EH, Korach KS, Eddy EM (2000) Spermatogenic cells do not require estrogen receptor-α for development of function. Endocrinology 141:1273–1276

Meistrich ML, van Beek MEAB (1993) Spermatogonial stem cells. In: Desjardins C, Ewing LL (eds) Cell and molecular biology of testis. Oxford University Press, Oxford, pp 266–295

Nagano M (2003) Homing efficiency and proliferation kinetics of male germ line stem cells following transplantation in mice. Biol Reprod 69:701–707

Nagano M, Avarbock MR, Leonida EB, Brinster CJ, Brinster RL (1998) Culture of mouse spermatogonial stem cells. Tissue Cell 30:389–397

Nagano M, Avarbock MR, Brinster RL (1999) Pattern and kinetics of mouse donor spermatogonial stem cell colonization in recipient testes. Biol Reprod 60:1429–1436

Nagano M, Brinster CJ, Orwig KE, Ryu B-Y, Avarbock MR, Brinster RL (2001a) Transgenic mice produced by retroviral transduction of male germ-line stem cells. Proc Natl Acad Sci U S A 98:13090–13095

Nagano M, McCarrey JR, Brinster RL (2001b) Primate spermatogonial stem cells colonize mouse testes. Biol Reprod 64:1409–1416

Nagano M, Patrizio P, Brinster RL (2002a) Long-term survival of human spermatogonial stem cells in mouse testes. Fertil Steril 78:1225–1233

Nagano M, Watson DJ, Ryu B-Y, Wolfe JH, Brinster RL (2002b) Lentiviral vector transduction of male germ line stem cells in mice. FEBS Lett 524:111–115

Nagano M, Ryu B-Y, Brinster CJ, Avarbock MR, Brinster RL (2003) Maintenance of mouse male germ line stem cells in vitro. Biol Reprod 68:2207–2214

Noguchi J, Toyama Y, Yuasa S, Kikuchi K, Kaneko M (2002) Hereditary defects in both germ cells and the blood-testis barrier system in as-mutant rats: evidence from spermatogonial transplantation and tracer permeability analysis. Biol Reprod 67:880–888

Ogawa T, Arechaga JM, Avarbock MR, Brinster RL (1997) Transplantation of testis germinal cells into mouse seminiferous tubules. Int J Dev Biol 41:111–122

Ogawa T, Dobrinski I, Avarbock MR, Brinster RL (1998) Leuprolide, a gonadotropin releasing hormone agonist, enhances colonization after spermatogonial transplantation into mouse testes. Tissue Cell 30:583–588

Ogawa T, Dobrinski I, Avarbock MR, Brinster RL (1999a) Xenogeneic spermatogenesis following transplantation of hamster germ cells to mouse testes. Biol Reprod 60:515–521

Ogawa T, Dobrinski I, Brinster RL (1999b) Recipient preparation is critical for spermatogonial transplantation in the rat. Tissue Cell 31:461–472

Ogawa T, Dobrinski I, Avarbock MR, Brinster RL (2000) Transplantation of male germ line stem cells restores fertility in infertile mice. Nature Med 6:29–34

Ogawa T, Ohmura M, Yumura Y, Sawada H, Kubota Y (2003) Expansion of murine spermatogonial stem cells through serial transplantation. Biol Reprod 68:316–322

Ohta H, Yomogida K, Dohmae K, Nishimune Y (2000) Regulation of proliferation and differentiation in spermatogonial stem cells: the role of c-kit and its ligand SCF. Development 127:2125–2131

Ohta H, Yomogida K, Tadokoro Y, Tohda A, Dohmae K, Nishimune Y (2001) Defect in germ cells, not supporting cells, is the cause of male infertility in the jsd mutant mouse: proliferation of spermatogonial stem cells without differentiation. Int J Androl 24:15–23

Orwig KE, Avarbock MR, Brinster RL (2002a) Retrovirus-mediated modification of male germline stem cells in rats. Biol Reprod 67:874–879

Orwig KE, Ryu B-Y, Avarbock MR, Brinster RL (2002b) Male germ-line stem cell potential is predicted by morphology of cells in neonatal rat testes. Proc Natl Acad Sci U S A 99:11706–11711

Parreira GC, Ogawa T, Avarbock MR, Franca LR, Brinster RL, Russell LD (1998) Development of germ cell transplants in mice. Biol Reprod 59:1360–1370

Radford J (2003) Restoration of fertility after treatment for cancer. Horm Res 59:21–23

Radford JA, Shalet SM, Lieberman BA (1999). Fertility after treatment for cancer. BMJ 319:935–936

Reis MM, Tsai MC, Schlegel PN, Feliciano M, Raffaelli R, Rosenwaks Z, Palermo GD (2000) Xenogeneic transplantation of human spermatogonia. Zygote 8:97–105

Rilianawati, Speed R, Taggart M, Cooke HJ (2003) Spermatogenesis in testes of Dazl null mice after transplantation of wild-type germ cells. Reproduction 126:599–604

Russell LD, Brinster RL (1996) Ultrastructural observations of spermatogenesis following transplantation of rat testis cells into mouse seminiferous tubules. J Androl 17:615–627

Russell LD, Ettlin RA, SinhaHikim AP, Clegg ED (1990) Mammalian spermatogenesis. In: Russell LD, Ettlin RA, SinhaHikim AP, Clegg ED (eds) Histological and histopathological evaluation of the testis. Cache River Press, Clearwater, pp 1–40

Ryu B-Y, Orwig KE, Avarbock MR, Brinster RL (2003) Stem cell and niche development in the postnatal rat testis. Dev Biol 263: 253–263

Ryu B-Y, Orwig KE, Kubota H, Avarbock MR, Brinster RL (2004) Phenotypic and functional characteristics of spermatogonial stem cells in rats. Dev Biol 274:158–170

Schlatt S, Foppiani L, Rolf C, Weinbauer GF, Nieschlag E (2002a) Germ cell transplantation into X-irradiated monkey testes. Hum Reprod 17:55–62

Schlatt S, Kim SS, Gosden R (2002b) Spermatogenesis and steroidogenesis in mouse, hamster and monkey testicular tissue after cryopreservation and heterotopic grafting to castrated hosts. Reproduction 124:339–346

Schlatt S, Honaramooz A, Boiani M, Scholer HR, Dobrinski I (2003) Progeny from sperm obtained after ectopic grafting of neonatal mouse testes. Biol Reprod 68:2331–2335

Shinohara T, Avarbock MR, Brinster RL (1999) β_1- and α_6-integrin are surface markers on mouse spermatogonial stem cells. Proc Natl Acad Sci U S A 96:5504–5509

Shinohara T, Avarbock MR, Brinster RL (2000a) Functional analysis of spermatogonial stem cells in Steel and cryptorchid infertile mouse models. Dev Biol 220:401–411

Shinohara T, Orwig KE, Avarbock MR, Brinster RL (2000b) Spermatogonial stem cell enrichment by multiparameter selection of mouse testis cells. Proc Natl Acad Sci U S A 97:8346–8351

Shinohara T, Orwig KE, Avarbock MR, Brinster RL (2001) Remodeling of the postnatal mouse testis is accompanied by dramatic changes in stem cell number and niche accessibility. Proc Natl Acad Sci U S A 98:6186–6191

Shinohara T, Inoue K, Ogonuki N, Kanatsu-Shinohara M, Miki H, Nakata K, Kurome M, Nagashima H, Toyokuni S, Kogishi K, Honjo T, Ogura A (2002) Birth of offspring following transplantation of cryopreserved immature testicular pieces and in-vitro microinsemination. Hum Reprod 17:3039–3045

Shinohara T, Orwig KE, Avarbock MR, Brinster RL (2003) Restoration of spermatogenesis in infertile mice by Sertoli cell transplantation. Biol Reprod 68:1064–1071

Tegelenbosch RAJ, de Rooij DG (1993) A quantitative study of spermatogonial multiplication and stem cell renewal in the C3H/101 F_1 hybrid mouse. Mut Res 290:193–200

Van Pelt MM, Roepers-Gajadien HL, Gademan IS, Creemers LB, de Rooij DG, van Dissel-Emiliani FMF (2002) Establishment of cell lines with rat spermatogonial stem cell characteristics. Endocrinology 143:1845–1850

Ventela S, Ohta H, Parvinen M, Nishimune Y (2002) Development of stages of the cycle in mouse seminiferous epithelium after transplantation of green fluorescent protein-labeled spermatogonial stem cells. Biol Reprod 66:1422–1429

Wistuba J, Schlatt S, Cantauw C, von Schonfeldt V, Nieschlag E, Behr R (2002) Transplantation of wild-type spermatogonia leads to complete spermatogenesis in testes of cyclic 3′,5′-adenosine monophosphate response element modulator-deficient mice. Biol Reprod 67:1052–1057

Yomogida K, Yagura Y, Tadokoro Y, Nishimune Y (2003) Dramatic expansion of germinal stem cells by ectopically expressed human glial cell line-derived neurotrophic factor in mouse Sertoli cells. Biol Reprod 69:1303–1307

Zhang Z, Renfree MB, Short RV (2003) Successful intra- and interspecific male germ cell transplantation in the rat. Biol Reprod 68:961–967

13 Progenitor Cell-Based Myelination as a Model for Cell-Based Therapy of the Central Nervous System

S.A. Goldman, J. Lang, N. Roy, S.J. Schanz, F.S. Sim, S. Wang,
V. Washco, M.S. Windrem

13.1   Introduction: Glial Progenitor Cells of the Adult Brain 196
13.2   Glial Progenitor Cell-Based Therapy
        of Adult Demyelinating Disease   200
13.3   The Childhood Disorders of Myelin as Therapeutic Targets 200
13.4   Neonatal Delivery of Oligodendrocyte Progenitor Cells
        as a Treatment for Congenital Leukodystrophies 201
13.5   Defining Optimal Cellular Vectors for the Leukodystrophies 202
13.6   Cell-Based Strategies for Treating Lysosomal Storage Disorders
        and Metabolic Leukodystrophies 207
13.7   Overview 209
References 210

Abstract. Diseases of the brain and spinal cord are especially daunting challenges for cell-based strategies of repair, given the multiplicity of cell types within the adult central nervous system, and the precision with which they must interact in both space and time. Nonetheless, a number of diseases are especially appropriate for cell-based therapy, in particular those in which single phenotypes are lost. Foremost among these are the disorders of myelin, in which oligodendrocytes are the specific and often sole victims of the underlying disease process. These include not only the vascular, traumatic, and inflammatory demyelinations of adulthood, but also the congenital and childhood dysmyelinating syndromes of the pediatric leukodystrophies. These congenital disorders of myelin formation and maintenance may present especially compelling targets for cell-based neurological therapy.

13.1   Introduction: Glial Progenitor Cells of the Adult Brain

Progenitor cells of the forebrain parenchyma very likely constitute the most abundant neural progenitor phenotypes of the adult human brain (Nunes et al. 2003; Scolding et al. 1998). These cells are nominally glial progenitors, and have been described as such across mammalian phylogeny. They are widely distributed in the adult brain and may comprise as many as 3%–4% of all cells in the adult human white matter (Roy et al. 1999, 2004). Competent glial progenitor cells may be isolated and harvested in bulk from human brain tissue, of both fetal and adult origin (Windrem et al. 2004). Although they have typically been described as bipotential astrocyte-oligodendrocyte glial progenitor cells, and are readily biased to an oligodendrocyte phenotype, at least some fraction can normally generate neurons as well (Belachew et al. 2003; Nunes et al. 2003). As such, these cells appear to act as tissue-restricted multipotential progenitors, restricted to glial phenotype by their environment, rather than by cell autonomous constraints. Such nominally glial progenitor cells may therefore actually comprise a pool of transit-amplifying multipotential progenitors of adult brain parenchyma, having translocated from the stem cell niche of the ventricular zone into the highly gliogenic niche of the adult white matter (Goldman 2003; Nunes et al. 2003; Sim et al. 2005) (Fig. 1).

Given the fundamentally gliogenic nature of these cells in vivo and their ready capacity to generate oligodendrocytes both in vivo and in vitro, they have been assessed as transplant vectors in a variety of experimental models of dysmyelinating diseases (reviewed in Duncan 2005; Goldman 2005; Keirstead 2005; Roy et al. 2004). To this end, methods have been developed by which these cells may be extracted to purity, from both fetal and adult tissues, as well as from human embryonic stem cells, using both promoter-specified GFP-based (Roy et al. 1999), and surface antigen-based (Dietrich et al. 2002; Windrem et al. 2002, 2004)

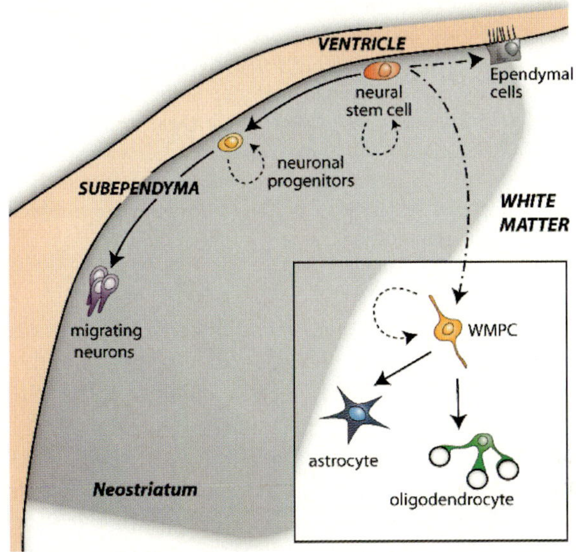

Fig. 1. Stem and progenitor cells of the adult human brain. This cartoon illustrates the basic categories of progenitor cells in the adult brain and their lineal relationships. The striatal wall of the forebrain is schematized here; it includes periventricular neural stem cells (*red*), which generate two major populations of transit-amplifying progenitors that contribute to neuronal and glial lineages (*orange*), respectively. These include the neuronal progenitor cells of the ventricular subependyma and the nominally oligodendrocyte progenitor cells of the subcortical white matter, designated here as white matter progenitor cells (*WMPCs*). Each of these transit-amplifying pools has been isolated from adult human brain (Roy et al. 1999, 2000a, 2000b), and each gives rise to differentiated progeny appropriate to its location, including neurons (*purple*), oligodendrocytes (*green*), and parenchymal astrocytes (*blue*). The neuronal migrants within the ventricular subependymal transit to the olfactory bulb in rodents (Alvarez-Buylla and Garcia-Verdugo 2002; Lois and Alvarez-Buylla 1994; Luskin 1993), though whether they do so in humans remains unclear (Sanai et al. 2004)

fluorescence-activated cell sorting (Fig. 2). However derived, these glial progenitor cells have been colloquially designated oligodendrocyte progenitor cells (Levine et al. 2001), and we will use this nomenclature here

Fig. 2. Sources, isolation, and use of defined progenitor phenotypes. This figure schematizes the methods of isolating prospectively defined neural progenitor phenotypes from a variety of human cell and tissue sources, and highlights several of the experimental purposes to which these cells may be allocated and devoted. Oligodendrocyte progenitors may be specifically isolated from the adult white matter or fetal subventricular zone, using either A2B5-based immunoextraction or CNP2 promoter-directed, GFP-based fluorescence-activated cell sorting (*FACS*). (Adapted from Goldman 2005)

as well, recognizing that these cells actually retain far broader latitude for generating other neural phenotypes, depending upon the environment to which they are exposed (Kondo and Raff 2000; Nunes et al. 2003).

Oligodendrocytes are the sole source of myelin in the adult CNS, and their loss or dysfunction is at the heart of a wide variety of diseases of both children and adults (Keirstead 2005). The demyelinating diseases are thus especially attractive targets for cell-based therapeutic strategies, since they are in effect caused by the loss of that single, relatively homogeneous phenotype, the myelinating oligodendrocyte. By virtue of their ability to readily generate these cells, oligodendrocyte progenitor cells (OPCs) have become a focus of study for those interested in restoring myelin to demyelinated regions of the diseased or injured CNS. Indeed, by virtue of their widespread and abundant distribution, OPCs may also prove intriguing targets for inductive strategies, intended to mobilize and utilize endogenous progenitor cell pools (Goldman 2004). In diseases with multifocal transient demyelination, such as occurs in MS, the ability to pharmacologically mobilize endogenous progenitor pools and direct their daughters to oligodendrocyte lineage might have great utility. Yet in most diseases of the brain and spinal cord, resident OPCs are themselves either lost – as in stroke and major injury – or diseased – as in the hereditary and metabolic leukodystrophies. In these cases, OPCs may be most useful as transplantable cellular vectors for remyelination. In this chapter, we shall focus on the latter set of efforts, intended to develop the use of isolated human OPCs as transplantable agents for mediating therapeutic remyelination.

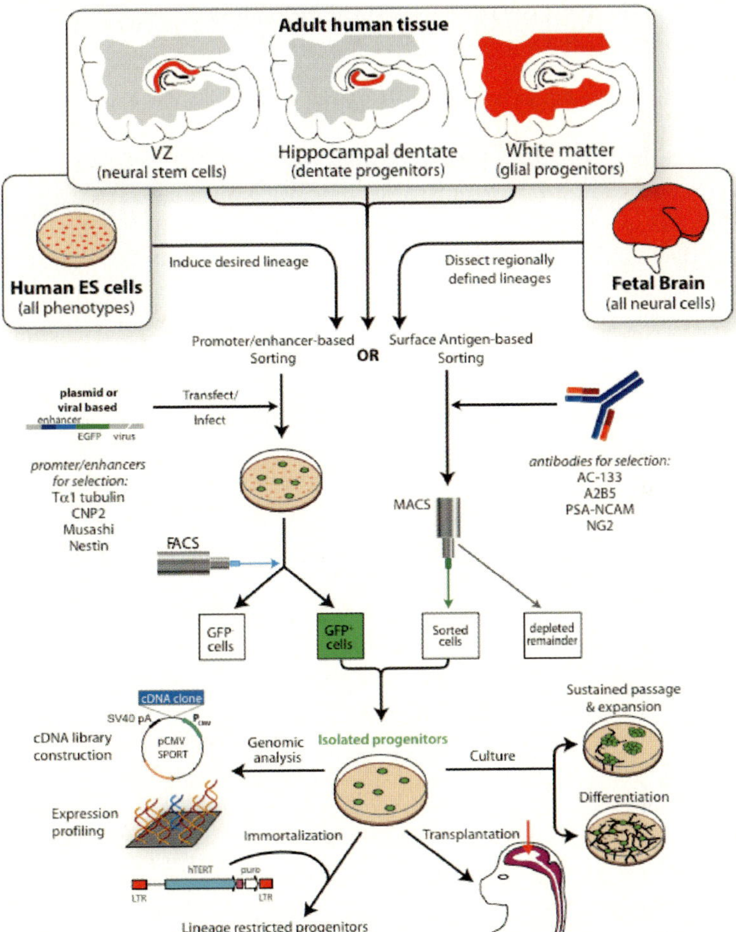
Adult human tissue
VZ
(neural stem cells)
Hippocampal dentate
(dentate progenitors)
White matter
(glial progenitors)
Human ES cells
(all phenotypes)
Fetal Brain
(all neural cells)
Induce desired lineage
Dissect regionally
defined lineages
Promoter/enhancer-based
Sorting
OR
Surface Antigen-based
Sorting
plasmid or
viral based
enhancer
EGFP   virus
Transfect/
Infect
promter/enhancers
for selection:
Tα1 tubulin
CNP2
Musashi
Nestin
FACS
antibodies for selection:
AC-133
A2B5
PSA-NCAM
NG2
MACS
GFP
cells
GFP+
cells
Sorted
cells
depleted
remainder
cDNA clone
SV40 pA   P CMV
pCMV
SPORT
cDNA library
construction
Genomic
analysis
Isolated progenitors
Culture
Sustained passage
& expansion
Expression
profiling
Immortalization
Transplantation
Differentiation
hTERT   puro
LTR   LTR
Lineage restricted progenitors
Experimental xenograft

13.2 Glial Progenitor Cell-Based Therapy of Adult Demyelinating Disease

In adults, oligodendrocytic loss is causal in diseases as diverse as the vascular leukoencephalopathies and multiple sclerosis (Roy et al. 2004). As a result, the engraftment of glial and oligodendrocyte progenitor cells has been assessed in a variety of models of demyelination in the adult brain and spinal cord. When transplanted into lysolecithin-lesioned adult rat brain, adult human OPCs were able to quickly mature as oligodendrocytes and myelinate residual denuded host axons, but with relatively low efficiency compared to the robust myelination seen in congenitally hypomyelinated brain (Windrem et al. 2002). Similarly, systemic administration of neural stem cells into mice subjected to experimental allergic encephalomyelitis resulted in some degree of local engraftment, oligodendrocytic maturation and myelination (Pluchino et al. 2003), although the robustness and stability of donor cell-mediated remyelination remains unclear (Pluchino et al. 2004). These observations highlight the importance of the disease environment in permitting appropriate oligodendrocytic differentiation and myelination (Franklin 2002). Thus, while human OPCs would seem effective agents by which to remyelinate acutely demyelinated brain tissue, the complexity of the adult disease environment may make such targets less imminently approachable than their pediatric counterparts, as discussed in Sect. 13.3. At the very least, cell-based therapeutic strategies for adult demyelination, especially those intended to remyelinate the inflammatory lesions of multiple sclerosis, will require aggressive disease modification and immunosuppression as adjuncts to cell delivery.

13.3 The Childhood Disorders of Myelin as Therapeutic Targets

The early dysmyelinations of the pediatric leukodystrophies comprise especially appropriate targets for progenitor cell-based therapy. Children suffer from a variety of hereditary diseases of myelin failure or loss, which include: (1) the hypomyelinating diseases, such as Pelizaeus-Merzbacher disease and hereditary spastic paraplegia, which are primary disorders of myelin formation; (2) the metabolic demyelinations and lysosomal storage disorders, such as metachromatic leukodystrophy, Tay-Sachs, Sandhoff's, and Krabbe's diseases, as well as adrenoleukodystrophy; and (3) gross myelinoclastic disorders of white matter loss, such as vanishing white matter disease and Canavan's disease, in which oligodendrocytes are early targets (reviewed in Kaye 2001; Powers 2004; van der Knapp et al. 2006). In addition, a variety of hereditary metabolic disorders that are manifested by early neuronal loss, such as the organic acidurias and the neuronal ceroid lipofuscinoses, are accompanied by early oligodendrocyte loss. Furthermore, periventricular leukomalacia, the most common single form of cerebral palsy, may be due in part to a perinatal loss of oligodendrocytes and their precursors (Back and Rivkees 2004; Follett et al. 2004; Robinson et al. 2005). Their mechanistic heterogeneity notwithstanding, all of these conditions include the prominent loss of oligodendrocytes and central myelin, highlighting the potential importance of restoring oligodendrocyte progenitor cells throughout this wide spectrum of perinatal disorders.

13.4 Neonatal Delivery of Oligodendrocyte Progenitor Cells as a Treatment for Congenital Leukodystrophies

To assess the potential of cell-based treatment for congenital dysmyelination, Windrem et al. transplanted sorted human oligodendrocyte progenitor cells (OPCs) of both fetal and adult origin into newborn shiverer mice, a dysmyelinated mouse deficient in myelin basic protein (Windrem et al. 2002, 2003, 2004). The donor OPCs widely dispersed throughout the white matter throughout the shiverer forebrain, such that single neonatal injections of OPCs into the lateral ventricles and adjacent callosum yielded abundant donor cell infiltration of the entire corpus callosum, fimbria, and internal and external capsules, as well as the deep subcapsular white matter to the level of the cerebral peduncles (Fig. 3a). Although the dorsal brainstem was not infiltrated by cells introduced to the forebrain, addition of a single intracerebellar injection at birth proved sufficient to substantially infiltrate the cerebellar white matter, peduncles, and dorsal brainstem, allowing donor engraftment contiguous with that of the forebrain and ventral brainstem (Fig. 3a, 3c–e). These cells migrated as widely as did both native and immortalized murine neural stem cells, which have been reported to be similarly capable of context-dependent differentiation and myelination (Mitome et al. 2001; Yandava et al. 1999).

Fig. 3. a–i Myelination by engrafted OPCs. **a** Implanted human fetal OPCs myelinated extensive regions of shiverer mouse forebrain. This animal was injected on P0 into the corpus callosum, cerebellar peduncles, and cisterna magnum with 1×10^5 cells at each site, then sacrificed at day 60 and stained for human nuclear antigen (*red*) to identify donor cells. **b** The striatocallosal border of a shiverer brain 3 months after engraftment with human fetal OPCs (hNA, *blue*). Donor-derived MBP (*red*) is evident in the callosum, while donor-derived GFAP$^+$ (*green*) astrocytes predominate in the striatum and along the ventricular wall. OPCs were thus recruited as oligodendrocytes or astrocytes in a context-dependent manner. **c–e** Extensive myelin basic protein (MBP) expression by sorted human OPCs, implanted into homozygote shiverer mice as neonates, indicates that large regions of the corpus callosum (**c, d**, different mice) have myelinated by 12 weeks (MBP, *green*). **e** OPCs myelinated fibers throughout the dorsoventral extent of the internal capsules (dorsal to *left*; ventral to *right*, extends to cerebral peduncles). **f** A confocal micrograph showing a triple-immunostain for MBP (*red*), human ANA (*blue*), and neurofilament protein (*green*). In this image, all MBP immunostaining is derived from the sorted human OPCs, whereas the NF$^+$ axons are those of the mouse host. *Arrows* identify murine axons ensheathed by human MBP. **g** A 2-µm-deep composite of optical sections, taken through the corpus callosum of a shiverer recipient sacrificed 12 weeks after fetal OPC implantation. Shiverer axons were scored as ensheathed when the *yellow* index lines intersected an NF$^+$ axon abutted on each side by MBP. The *asterisk* indicates the field enlarged in the inset (*right*). **h–i** Representative electron micrographs of 16-week-old shiverer homozygotes implanted with human OPCs shortly after birth. The images show shiverer axons ensheathed by densely compacted myelin. The *asterisk* in **i** indicates the field enlarged in the inset. Inset (*upper right*): Major dense lines are noted between lamellae, providing EM confirmation of myelination. **j** High-power confocal images of MBP$^+$ donor-derived myelin sheaths (*green*) at internodal junctions, characterized by expression of Caspr protein (*red*) at the paranodal borders. *Left*: A *z*-stack composite; *right:* a single 0.4-µm optical section. Caspr staining confirmed nodes of Ranvier between adjacent donor-derived myelinated segments (Einheber et al. 1997); these results suggest physiologically appropriate conduction support by donor-derived myelin. Scale = **f** 20 µm; **g** 40 µm, **h, i** 1 µm. (Adapted from Goldman 2005; Windrem et al. 2004)

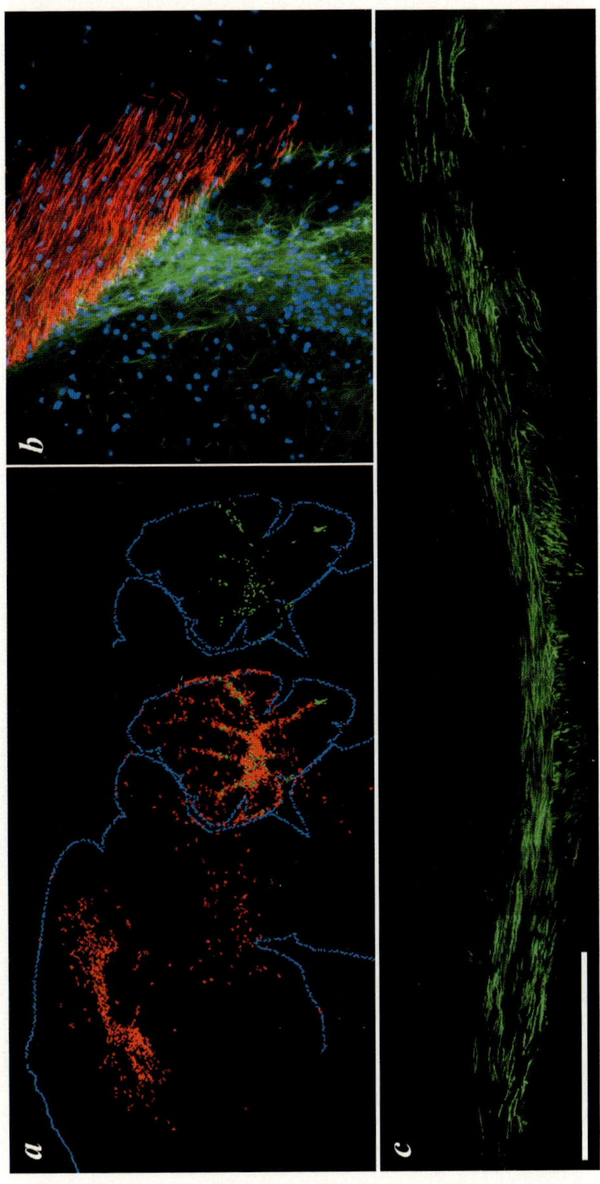

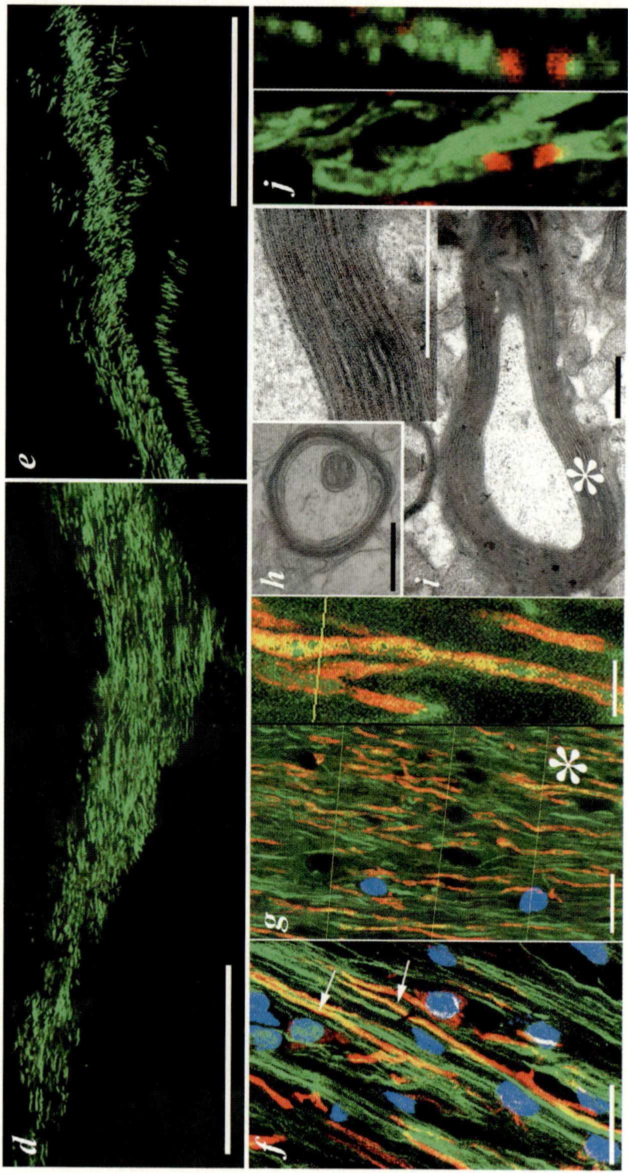

Neonatally implanted human donor OPCs developed as astrocytes and myelinating oligodendrocytes in a context-dependent fashion, such that those donor cells that engrafted presumptive white matter developed as oligodendrocytes, while those invading cortical and subcortical gray developed largely as astrocytes (Fig. 3b) (Windrem et al. 2004). The majority of donor cells engrafted the white matter, such that after neonatal intraventricular and intracallosal injection, the corpus callosum typically expressed MBP throughout its mediolateral extent, along its entire length in the sagittal plane, and throughout the dorsoventral extent of the internal capsules to the cerebral peduncles (Windrem et al. 2004) (Fig. 3a–c). Donor-derived myelin effectively ensheathed host shiverer axons, as validated by both confocal imaging (Fig. 3f, g) and the ultrastructural observation of donor-derived myelin with major dense lines (Fig. 3h, i), indicating effective myelin compaction, of which native shiverer oligodendrocytes are incapable. In addition, confocal analysis revealed the presence of nodes of Ranvier between donor-derived myelinated segments, and the paranodal expression of Caspr protein, suggesting functionally appropriate nodal architecture (Fig. 3j). Given the widespread dispersal of donor cells, their high-density engraftment and myelination and their architecturally-appropriate and quantitatively significant ensheathment of host axons, these results strongly suggested the feasibility of neonatal progenitor cell implantation as a potential therapeutic strategy in the congenital disorders of myelin.

13.5 Defining Optimal Cellular Vectors for the Leukodystrophies

Both fetal and adult OPCs have been proposed as potential cellular vectors for cell therapy of the congenitally dysmyelinated CNS. Yet fetal and adult-derived OPCs have been observed to behave quite differently after neonatal xenograft (Windrem et al. 2004). Isolates of human OPCs derived from adult white matter myelinated recipient brain much more rapidly than did fetal OPCs; adult-derived progenitors achieved widespread myelination by just 4 weeks after graft, while cells derived from late second-trimester fetuses took over 3 months to do so. The adult OPCs also generated oligodendrocytes more efficiently than fetal glial progenitors, and ensheathed more axons per donor cell than did fetal cells. In contrast, fetal glial progenitors emigrated more widely and engrafted more efficiently, differentiating as astrocytes in gray matter regions and oligodendrocytes in white matter (Fig. 4).

The divergent behaviors of fetal and adult-derived glial progenitors suggests their respective use for different disease targets. Adult OPCs, by virtue of their oligodendrocytic bias and rapid myelination, may be most appropriate for diseases of acute oligodendrocytic loss, such as subcortical infarcts and postinflammatory demyelinated lesions. In contrast, fetal progenitors may prove more effective for treating the congenital leukodystrophies, since their extensive emigration better assures uniform and widespread dispersal, while their astrocytic differentiation and invasion of grey matter may offer the correction of enzymatic deficits in deficient cortex.

Fig. 4. a–j Fetal and adult OPCs differed substantially in their speed and efficiency of myelinogenesis. **a** Adult-derived human OPCs (hNA, *red*) achieved dense MBP (*green*) by 4 weeks after graft. **b** In contrast, fetal OPCs expressed no MBP at 4 weeks, and none until 12 weeks (**c**). **c, d** Low- and high-power images of the callosal-fimbrial junction of a shiverer brain; dense myelination by 12 weeks after perinatal delivery of adult OPCs. **e, f** Adult OPCs developed mature myelin ultrastructure and major dense lines within 5 weeks of perinatal injection. **e, f** Myelin in a shiverer homozygote 5 weeks after perinatal injection of adult OPCs. Mice injected with fetal OPCs exhibited no evidence of myelination at this time point. **g** The distribution of hNA$^+$ adult OPCs (*red*), 4 weeks after implant. **h** A higher proportion of adult OPCs developed MBP expression then did fetal OPCs, when assessed 12 weeks after transplant. **i** Fetal OPCs nonetheless engrafted more efficiently and in higher numbers than did adult OPCs. * $p <0.05$; ** $p <0.005$, Student's *t*-test. **j** A plot comparing the number of ensheathed axons per donor cell achieved by fetal and adult-derived OPCs. Ensheathment was defined by confocal-imaged MBP$^+$ enwrapping NF$^+$ axons and was measured as a function of total donor cell number (*left panel*) and of MBP$^+$ donor-derived oligodendrocytes (*right panel*). The difference between fetal and adult donor ensheathment efficiencies was significant by Mann-Whitney ($p <0.02$). Scale: **a, b** 100 µm, **c** 1 mm; **d** 30 µm; **e, f** 1 µm. (Adapted from Windrem et al. 2004)

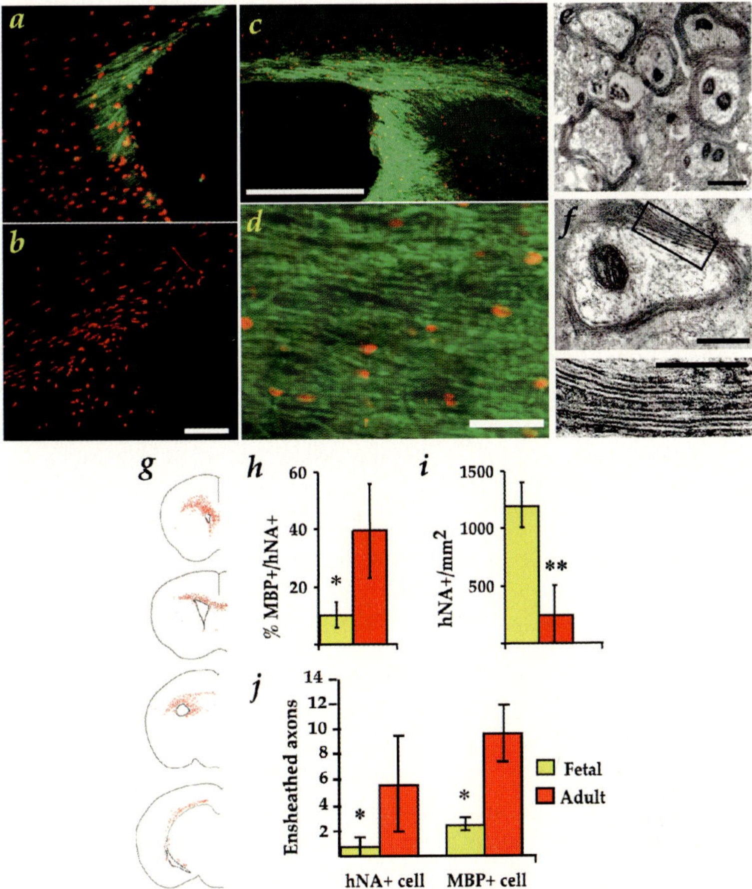

13.6 Cell-Based Strategies for Treating Lysosomal Storage Disorders and Metabolic Leukodystrophies

In the metabolic disorders of myelin, such as Krabbe's and Canavan's diseases, oligodendrocytes are essentially bystanders, killed by toxic metabolites emanating from cells deficient in one or more critical enzymes (Kaye 2001; Powers 2004; Suzuki 2003). Since the engraftment of glial progenitor cells is associated with astrocytic as well as oligoden-drocytic production, and since both the subcortical and cortical gray are infiltrated with donor-derived astrocytes after early implantation, glial progenitors would seem an especially promising vehicle for the distribution of enzyme-producing cells throughout otherwise deficient brain parenchyma. On that basis, several groups have begun to assess the ability of enzymatically competent, effectively wild-type glial progenitor cells to delay or ameliorate the signs and symptoms of the lysosomal storage disorders and other metabolic leukodystrophies. Indeed, perinatal grafts of fetal progenitor cells might prove a means of simultaneously myelinating and correcting enzymatic deficiencies in the pediatric leukodystrophies (Snyder et al. 1995). The lysosomal storage disorders present especially attractive targets in this regard, since wild-type lysosomal enzymes may be released by integrated donor cells and taken up by deficient host cells through the mannose-6-phosphate receptor pathway (Urayama et al. 2004). As a result, a relatively small number of donor glia may provide sufficient enzymatic activity to correct the underlying catalytic deficit and storage disorder of a much larger number of host cells (Jeyakumar et al. 2005).

Snyder and colleagues first established this principle of cell-based rescue of enzymatically deficient host cells by wild-type donor NSC implantation in a mouse model of Sly's disease (MPS-VII), in which myc-transduced NSCs were implanted neonatally and observed to migrate widely and restore lost enzymatic function broadly in the recipient forebrain (Snyder et al. 1995). The same group subsequently reported expression of β-hexosaminidase upon engraftment of transduced neural stem cells into recipient mice (Lacorazza 1996), though functional benefits accruing to engraftment-associated enzyme expression have not yet been reported. Similarly, Pellegatta et al. recently engrafted twitcher mice, a murine model of Krabbe's globoid cell leukodystrophy, with cultured neural stem cells transduced to over-express galactocerebrosidase, the enzyme deficient in Krabbe's disease (Pellegatta et al. 2005). Although the engrafted cells did not survive well in the highly inflammatory twitcher brain, they migrated appropriately to active sites of demyelination, in a manner akin to that noted in adults by Martino et al. (Pluchino et al. 2003, 2004). One might hope that in recipients immunosuppressed to reduce donor cell rejection, engrafted progenitors may indeed prove competent to prevent progressive demyelination in the lysosomal storage disorders and metabolic leukodystrophies.

That being said, few data currently exist with regards to the number or proportion of wild-type cells required to achieve local correction of enzymatic activity and substrate clearance in any storage disorder, and these values will likely need to be obtained for each disease target. Similarly, effective cell doses, delivery sites, and time frames will need to be established in models of congenital hypomyelination, before clinical trials of progenitor-based therapy can be contemplated. Moreover, the efficiency of myelination required for significant benefit remains unclear, as functional improvement may require remyelination over much if not the entire linear extent of each recipient axon. These caveats notwithstanding, there is reason for optimism that cell-based therapy of the pediatric myelin disorders, in particular for the primary dysmyelinations such as Pelizaeus-Merzbacher disease, vanishing white matter disease, and the spastic diplegic forms of cerebral palsy, may not be far off.

13.7   Overview

Clinically responsible stem-cell based therapy of the diseased CNS mandates a rigorous determination of those diseases most amenable to cell-based therapy. The glial diseases stand out as especially promising initial targets for progenitor cell therapy of CNS disease. Using a common strategy of glial progenitor cell implantation, pediatric diseases as diverse as the leukodystrophies, lysosomal storage diseases and cerebral palsy, and such adult acquired demyelinations as transverse myelitis, multiple sclerosis, and subcortical stroke, may all be approachable as therapeutic targets. Moreover, the same technologies used to enrich progenitor cells for transplantation yield isolates amenable to both immortalization and gene expression analyses as well (Sim et al. 2006). As a result, isolated glial progenitor cells may prove useful not only as cellular vectors for transplantation, but as tools for understanding the signaling pathways and growth control of native progenitors in vivo. Using such information, endogenous glial progenitor cells can be targeted for directed mobilization and phenotypic induction, whether by cognate cytokines, or by their small molecule mimetics. Indeed, by mobilizing endogenous progenitors in vivo, we can hope to mitigate the need for transplantation in disorders such as the ischemic and inflammatory demyelinations, in which large accessible stores of endogenous progenitors may persist locally. In contrast, in diseases involving the widespread loss of cells, and disorders in which endogenous progenitor cells themselves are lost or deficient, such as the congenital leukodystrophies and lysosomal storage disorders, therapeutic strategies based on cell transplantation will be necessary. Together, these distinct approaches highlight the potential utility of glial progenitor cell-based therapy in diseases of both the pediatric and adult central nervous system.

Acknowledgements. Dr. Goldman is supported by NINDS, the National Multiple Sclerosis Society, the New York Spinal Cord Injury Research Program, the A-T Children's Project and the CNS Foundation, Berlex Bioscience, and Q Therapeutics. Correspondence to Dr. Goldman: steven-goldman@urmc.rochester.edu.

References

Alvarez-Buylla A, Garcia-Verdugo JM (2002) Neurogenesis in adult subventricular zone. J Neurosci 22:629–634

Back S, Rivkees S (2004) Emerging concepts in periventricular white matter injury. Semin Perinatol 6:405–414

Belachew S, Chittajallu R, Aguirre AA, Yuan X, Kirby M, Anderson S, Gallo V (2003) Postnatal NG2 proteoglycan-expressing progenitor cells are intrinsically multipotent and generate functional neurons. J Cell Biol 161:169–186

Dietrich J, Noble M, Mayer-Proschel M (2002) Characterization of A2B5+ glial precursor cells from cryopreserved human fetal brain progenitor cells. 40:65–77

Duncan I (2005) Oligodendrocytes and stem cell transplantation: their potential in the treatment of leukoencephalopathies. J Inher Metab Dis 28:357–368

Einheber S, Zannazi G, Ching W, Scherer S, Milner T, Peles E, Salzer J (1997) The axonal membrane protein Caspr, a homologue of neurexin IV, is a component of the septate-like paranodal junctions that assemble during myelination. J Cell Biol 139:1495–1506

Follett P, Deng W, Dai W, Talos D, Massilon L, Rosenberg P, Volpe J, Jensen F (2004) Glutamate receptor-mediated oligodendrocyte toxicity in periventricular leukomalacia: a protective role for topiramate. J Neurosci Res 24:4412–4420

Franklin RJ (2002) Why does remyelination fail in multiple sclerosis? Nat Rev Neurosci 3:705–714

Goldman S (2003) Glia as neural progenitor cells. Trends Neurosci 26:590–596

Goldman S (2004) Directed mobilization of endogenous neural progenitor cells: the intersection of stem cell biology and gene therapy. Curr Opin Molecular Ther 6:466–472

Goldman SA (2005) Stem and progenitor cell-based therapy of the human central nervous system. Nat Biotech 23:862–871

Jeyakumar M, Dwek R, Butters T, Platt F (2005) Storage solutions: treating lysosomal disorders of the brain. Nat Rev Neurosci 6:1–12

Kaye E (2001) Update on genetic disorders affecting white matter. Pediatr Neurology 24:11–24

Keirstead H (2005) Stem cells for the treatment of myelin loss. Trends Neurosci 28:677–683

Kondo T, Raff M (2000) Oligodendrocyte precursor cells reprogrammed to become multipotential CNS stem cells. Science 289:1754–1757

Lacorazza HD (1996) Expression of human β-hexosaminidase α-subunit gene (the gene defect of Tay-Sachs disease) in mouse brains upon engraftment of transduced progenitor cells. Nat Med 2:424–429

Levine JM, Reynolds R, Fawcett JW (2001) The oligodendrocyte precursor cell in health and disease. Trends Neurosci 24:39–47

Lois C, Alvarez-Buylla A (1994) Long-distance neuronal migration in the adult mammalian brain. Science 264:1145–1148

Luskin MB (1993) Restricted proliferation and migration of postnatally generated neurons derived from the forebrain subventricular zone. Neuron 11:173–189

Mitome M, Low HP, van Den Pol A, Nunnari JJ, Wolf MK, Billings-Gagliardi S, Schwartz WJ (2001) Towards the reconstruction of central nervous system white matter using neural precursor cells. Brain 124:2147–2161

Nunes MC, Roy NS, Keyoung HM, Goodman RR, McKhann G, Jiang L, Kang J, Nedergaard M, Goldman SA (2003) Identification and isolation of multipotential neural progenitor cells from the subcortical white matter of the adult human brain. Nat Med 9:439–447

Pellegatta S, Tunici P, Poliani P, Dolcetta D, Cajola L, Colombelli C, Ciusani E, DiDonato S, Finocchiaro G (2006) The therapeutic potential of neural stem/progenitor cells in murine globoid cell leukodystrophy is conditioned by macrophage/microglial activation. Neurobiol Dis 21:314–323

Pluchino S, Furlan R, Martino G (2004) Cell-based remyelinating therapies in multiple sclerosis: evidence from experimental studies. Curr Opin Neurol 17:247–255

Pluchino S, Quattrini A, Brambilla E, Gritti A, Salani G, Dina G, Galli R, Del Carro U, Amadio S, Bergami A, Furlan R, Comi G, Vescovi AL, Martino G (2003) Injection of adult neurospheres induces recovery in a chronic model of multiple sclerosis. Nature 422:688–694

Powers J (2004) The leukodystrophies: overview and classification. In: Lazzarini RA (ed) Myelin biology and disorders, Vol. 2. Elsevier Academic Press, San Diego, pp 663–690.

Robinson S, Petelenz K, Li Q, Cohen M, Dechant A, Tabrizi N, Bucek M, Lust D, Miller R (2005) Developmental changes induced by graded prenatal systemic hypoxic-ischemic insults in rats. Neurobiol Dis 18:568–581

Roy NS, Wang S, Harrison-Restelli C, Benraiss A, Fraser RA, Gravel M, Braun PE, Goldman SA (1999) Identification, isolation, and promoter-defined separation of mitotic oligodendrocyte progenitor cells from the adult human subcortical white matter. J Neurosci 19:9986–9995

Roy NS, Benraiss A, Wang S, Fraser RA, Goodman R, Couldwell WT, Nedergaard M, Kawaguchi A, Okano H, Goldman SA (2000a) Promoter-targeted selection and isolation of neural progenitor cells from the adult human ventricular zone. J Neurosci Res 59:321–331

Roy NS, Wang S, Jiang L, Kang J, Benraiss A, Harrison-Restelli C, Fraser RA, Couldwell WT, Kawaguchi A, Okano H, Nedergaard M, Goldman SA (2000b) In vitro neurogenesis by progenitor cells isolated from the adult human hippocampus. Nat Med 6:271–277

Roy N, Windrem M, Goldman SA (2004) Progenitor cells of the adult white matter. In: Lazzarini RA (ed) Myelin biology and disorders, Vol. 2. Elsevier Academic Press, San Diego, pp 259–287

Sanai N, Tramontin AD, Quinones-Hinojosa A, Barbaro NM, Gupta N, Kunwar S, Lawton MT, McDermott MW, Parsa AT, Manuel-Garcia VJ, Berger MS, Alvarez-Buylla A (2004) Unique astrocyte ribbon in adult human brain contains neural stem cells but lacks chain migration. Nature 427:740–744

Scolding N, Franklin R, Stevens S, Heldin CH, Compston A, Newcombe J (1998) Oligodendrocyte progenitors are present in the normal adult human CNS in the lesions of multiple sclerosis [see comments]. Brain 121:2221–2228

Sim FJ, Lang JK, Waldau B, Roy NS, Schwartz T, Pilcher W, Chandross K, Natesan S, Merrill J, Goldman SA (2006) Complementary patterns of gene expression by human oligodendrocyte progenitors and their environment predict determinants of progenitor maintenance and differentiation. Annals Neurology 59: 763–779

Sim FJ, Goldman SA (2005) White matter progenitor cells reside in an oligoden-drogenic niche. In: Opportunities and Challenges of the Therapies Targeting CNS Regeneration. D. Perez, B. Mitrovic, A. Baron, van Evercooren (Eds.) Ernst Schering Research Foundation. Springer, Berlin pp. 61–82

Snyder EY, Taylor RM, Wolfe JH (1995) Neural progenitor cell engraftment corrects lysosomal storage throughout the MPSVII mouse brain. Nature 374:367–370

Suzuki K (2003) Globoid cell leukodystrophy (Krabbe's disease): update. J Child Neurol 18:595–603

Urayama A, Grubb J, Sly W, Banks W (2004) Developmentally regulated man-nose 6-phophate receptor-mediated transport of a lysosomal enzyme across the blood-brain barrier. Proc Natl Acad Sci U S A 101:12658–12663

van der Knaap M, Pronk JC, Scheper GC (2006) Vanishing white matter disease. Lancet Neurology 5:413–423

Windrem MS, Nunes MC, Rashbaum WK, Schwartz TH, Goodman RA, McK-hann G, Roy NS, Goldman SA (2004) Fetal and adult human oligoden-drocyte progenitor cell isolates myelinate the congenitally dysmyelinated brain. Nat Med 10:93–97

Windrem MS, Roy N, Nunes M, Goldman SA (2003) Identification, selection and use of adult human oligodendrocyte progenitor cells. In: Zigova Sny-der E (ed) Neural stem cells for brain repair. Humana, New York, pp 69–88

Windrem MS, Roy NS, Wang J, Nunes M, Benraiss A, Goodman R, Mc-Khann GM, Goldman SA (2002) Progenitor cells derived from the adult hu-man subcortical white matter disperse and differentiate as oligodendrocytes within demyelinated lesions of the rat brain. J Neurosci Res 69:966–975

Yandava BD, Billinghurst LL, Snyder EY (1999) "Global" cell replacement is feasible via neural stem cell transplantation: evidence from the dysmyeli-nated shiverer mouse brain. Proc Natl Acad Sci U S A 96:7029–7034

14 Adult Neural Stem Cells and Central Nervous System Repair

H. Okano

14.1   Introduction 216
14.1.1 The Concept of Regeneration of the Central Nervous System . . . 216
14.1.2 Normal Developmental Process
       of the Vertebrate Central Nervous System 216
14.2   Tools for Stem Cell Research in Relation
       to the Central Nervous System 218
14.3   Identification of Neural Stem/Progenitor Cells
       in the Adult Human Brain 219
14.4   Insult-Induced Neurogenesis 219
14.5   The RNA-Binding Protein Musashi-1
       as a Stem Cell Marker and Its Functions 221
14.6   Blockade of Interleukin-6 Signaling in Spinal Cord Injury 222
References 225

Abstract. It has long been believed that the adult mammalian central nervous system does not regenerate after injury. However, recent advances in the field of stem cell biology, including the identification of Musashi-1-positive neural stem cells (NSCs) or NSC-like cells, has provided new insight for the development of novel therapeutic strategies aimed at inducing regeneration in the damaged central nervous system (CNS). The major strategies for inducing regeneration in the damaged CNS can be classified into two subgroups: (1) activation of endogenous neural stem cells and (2) cell transplantation therapies. In this paper, we would like to summarize our recent findings on the functions of the neural RNA-binding protein Musashi-1 expressed in neural stem cells in relation to insult-induced neurogenesis, and therapeutic interventions for spinal cord injury, especially focusing on the treatment of spinal cord injury in the acute phase with anti-IL-6 receptor blocking antibody.

14.1 Introduction

14.1.1 The Concept of Regeneration of the Central Nervous System

For many years, ever since it was first pointed out by Ramon y Cajal in 1928, it has been believed that neural cells in the adult mammalian central nervous system (CNS) cannot regenerate after injury (Ramón y Cajal 1928). With the recent advances in stem cell technology and other methods, we believe that we might be able to challenge this long-held contention. However, before going into further detail, I would like to clarify the concepts of regeneration of the CNS (Fig. 1). While most neuroscientists classically define the term "neural regeneration" as "axonal regeneration", i.e., *re*growth of disrupted neuronal axons, *re*plenishment of neural stem cells lost as a result of injury or disease is increasingly being recognized as another important aspect of neural regeneration. Furthermore, from the point of view of therapeutic application, it is also important to achieve *re*covery of the lost neural functions. Thus, I believe that these are the three crucial aspects of regeneration of the CNS (Okano 2003). Then how can regeneration of the CNS be achieved? At this point, I believe that it is important to *re*capitulate the processes involved in normal CNS development.

14.1.2 Normal Developmental Process of the Vertebrate Central Nervous System

Figure 2 shows the normal developmental process of the CNS in vertebrate animals. The process is classified into major two stages, i.e., the neuronal-activity-independent earlier stage and the neuronal-activity-dependent later stage. Notably, the earliest event in CNS development is the induction of neuroepithelial cells, including neural stem cells

<u>Regeneration of the Central Nervous System</u>

Regrowth of disrupted neuronal axons

Replenishment of lost neuronal (or neural) cells

Recovery of lost neuronal (or neural) functions

How can the regeneration of CNS be achieved?

Recapitulate some aspects of normal CNS development

Fig. 1. The concept of regeneration of the CNS

Neuronal Activity-independent Early Stages
        -Nature-
- I.   **Induction of Neuroepithelial Cells (Neural Stem Cells) and Patterning.**
- II. **Production of Neurons and Glia from Neural Stem Cells and their subsequent migration.**
- III. **Decision of specific Cell Fates.**
- IV. **Guidance of Neuronal Growth Cone.**
- V.   **Synapse Formation**

Neuronal Activity-dependent Late Stages.
        -Nature-
- VI. **Cell Deaths regulated by Neutrotrophic Factors.**
- VII. **Synaptic Rearrangements acquired by competition and neuronal activity.**
- VII. **Synaptic Plasticity that takes place all through the animals' life.**

Fig. 2. The normal developmental process of the vertebrate CNS. Modified from the original ideas described by Goodman and Doe (1993)

(NSCs), from undifferentiated ectodermal cells. This served as the rationale for our utilizing and/or mobilizing NSCs in order to recapitulate normal CNS development.

14.2 Tools for Stem Cell Research in Relation to the Central Nervous System

NSCs belong to the category of tissue-specific stem cells and are present within the CNS. They are characterized by multi-lineage potential and self-renewal activity (Okano 2002a, b). Recently, there has been rapid progress in the area of stem cell biology of the CNS. I believe that one of the major reasons for such progress has been the development of tools for NSC research, including the development of (1) selective markers, (2) selective in vitro culture methods, and (3) methods for prospective identification and isolation of NSCs (Fig. 3). The selective marker molecules listed in Fig. 3 are not exclusively specific to NSCs; however, they are strongly expressed in NSCs, at least within the CNS.

Musashi-1 is an RNA-biding protein that has been shown to be preferentially expressed in neural precursor cells, including NSCs (Sakakibara et al. 1996; Sakakibara and Okano 1997; Kaneko et al. 2000). The functions of this molecule are described in a later section of this article. With regard to in vitro culture methods of NSCs, the neurosphere method is currently being used as the standard technology, in view of the ease with which they can be cultured in vitro. In terms of prospective identification of NSCs, many researchers have investigated cell-surface-based methods (Uchida et al. 2000; Rietze et al. 2001) similar to those used

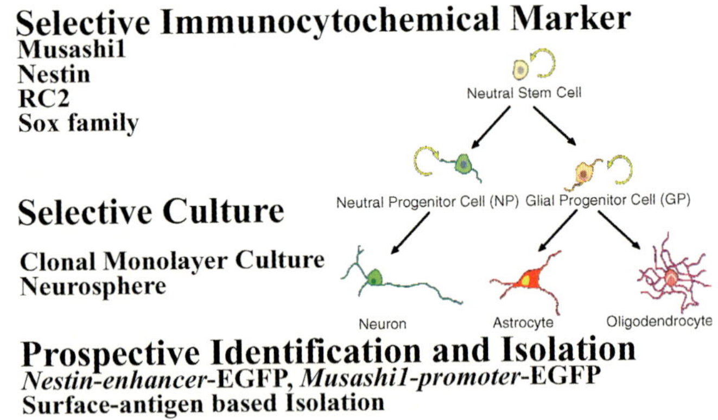

Fig. 3. Tools for neural stem cell (NSC) research

for the identification of hematopoietic stem cells (Osawa et al. 1996; Matsuzaki et al.). On the other hand, we use a different approach; in order to visualize the activity of the NSCs in situ and to isolate viable NSCs, we have constructed NSC-selective GFP-reporter genes, including nestin 2nd intronic enhancer-driven GFP (*nestin*-GFP) (Roy et al. 2000; Kawaguchi et al. 2001) and Musashi-1 promoter- driven GFP (*Musashi-1*-GFP) (Keyoung et al. 2001).

14.3   Identification of Neural Stem/Progenitor Cells in the Adult Human Brain

Using the above-mentioned tools, we attempted to identify the neural stem/progenitor cells in the adult human brain, in collaboration with the group of Dr. Steve Goldman in the United States. We examined Musashi-1-immunostained sections of the adult human brain of a 27-year-old man (Pincus et al. 1998). Scattered Musashi-1-positive cells were prominently identified in the periventricular area. Even though the immunohistochemical staining was conducted on fixed sections, these Musashi-1-positive cells are considered likely to represent neural stem/progenitor activities (Pincus et al. 1998). These putative neural stem/progenitor cells were isolated in a viable state using GFP-reporter genes. The isolated cells were cultured in the presence of FGF-2 and gave rise to functionally active neurons in vitro (Roy et al. 2000). These findings indicate that adult human brains do contain NSCs or NSC-like cells in the periventricular area. Based on this finding, we have proposed two major strategies for inducing regeneration of the damaged CNS using NSCs, i.e., (1) activation of endogenous NSCs and (2) cell therapy, including transplantation of NSCs. In this paper, I shall focus on the former strategy, while the details of cell therapy will be described elsewhere (Sawamoto et al. 2001; Ogawa et al. 2002; Okano et al. 2003; Iwanami et al. 2005a, b).

14.4   Insult-Induced Neurogenesis

Many researchers have identified NSCs or NSC-like cells in the adult mammalian CNS along the entire neuroaxis from the forebrain to the spinal cord. However, despite the presence of endogenous NSCs, most parts of the adult mammalian CNS are still non-neurogenic, which means that new neurons are not generated. Nonetheless, there are two well-known exceptions, the so-called neurogenic regions; the neurogenic regions in the adult mammalian brain are the (1) anterior part of the subventricular zones facing the lateral ventricle, which produces the olfactory interneurons, and (2) the hippocampal dentate gyrus (reviewed by Temple and Alvarez-Buylla 1999). In both of these regions, adult neurogenesis takes place under physiologic conditions and has physiologic significance (Kemperman and Gage 1999). It is noteworthy that the striatum and cerebral cortex do not show any detectable adult neurogenesis under physiologic conditions, even though they are major targets of brain ischemia. However, recent studies have shown that some insults, including ischemia, can induce neurogenesis in these regions of the brain (Arvidsson et al. 2002; Tonchev et al. 2005). Such neurogenesis induced by pathogenic conditions is called insult-induced neurogenesis. The detection of insult-induced neurogenesis clearly shows that some regenerative capacity is preserved in the adult mammalian brain, which challenges Cajal's dogma.

However, in spite of the detection of insult-induced neurogenesis, we must admit, nevertheless, that this regenerative capacity of the adult mammalian CNS has severe limitations. Why is this so? The potential problems associated with insult-induced neurogenesis may be summarized as follows:

1. The efficiency is low.
2. The newly generated neurons in response to CNS insults are mostly short-lived, possibly due to the absence of formation of any functional synapses.
3. In most cases, insult-induced neurogenesis per se is not sufficient for functional recovery of the neural deficits caused by the insults.
4. In the adult spinal cord, even insult-induced neurogenesis does not take place.

How can we overcome these difficulties? Adult neurogenesis is characterized by sequential events, as follows: (1) activation of adult NSCs, (2) proliferation of transitional amplifying cells and migration of neuroblasts, and (3) maturation of the newly born neurons (Lindvall et al.

2004). Thus, in order to improve the efficiency of insult-induced neurogenesis, it is essential to investigate the mechanisms underlying the above-mentioned events. We have proposed several approaches to address these questions. Here, we would like to introduce some of these approaches.

14.5 The RNA-Binding Protein Musashi-1 as a Stem Cell Marker and Its Functions

In order to enhance the efficiency of insult-induced neurogenesis, it is important to elucidate the mechanisms regulating the maintenance and activation of NSCs in the adult mammalian brain. We therefore focused our attention on Musashi-1, an RNA-binding protein that is strongly expressed in the mammalian NSCs (Sakakibara et al. 1996; Sakakibara and Okano 1997; Kaneko et al. 2000). Musashi constitutes an evolutionarily conserved family of RNA-binding proteins (Okano et al. 2002, 2005). Musashi-1 was shown to bind *m-Numb* mRNA in vivo and inhibit its translation, thereby augmenting Notch signaling (Imai et al. 2001). Many studies have indicated that Notch signaling plays essential roles in the self-renewal of NSCs (Nakamura et al. 2000; Kohyama et al. 2005). In fact, we observed the reduced self-renewal activity of NSCs in loss-of-function studies of Musashi. The functions of Musashi-1 are summarized in Fig. 4.

How then is the expression of Musashi-1 regulated? In the promoter region of the Musashi-1 gene, candidate sequences that can bind transcription factors involved in stem cell maintenance have been identified (unpublished observation). Interestingly, we found the presence of clusters of hypoxia-responsive element (HRE)-like sequences (ACGTG) in the promoter region of Musashi-1, which is consistent with the observation that Musashi-1 expression is upregulated in the ischemic brain (Yagita et al. 2002). Thus, one of the attractive hypotheses to explain the mechanism underlying insult-induced neurogenesis would be that hypoxia-induced Musashi-1 expression activates the endogenous NSCs through the translation repression of m-Numb and the subsequent activation of Notch signaling.

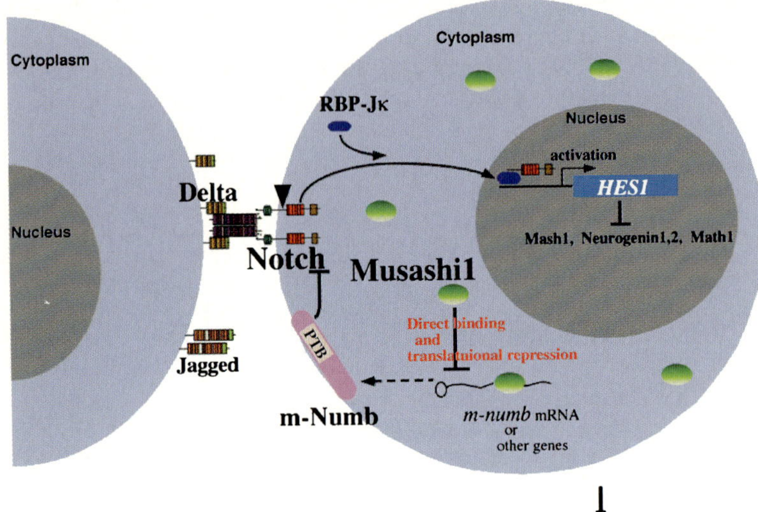

Fig. 4. Mechanism of involvement of Musashi-1 in the self-renewal of NSCs. Musashi-1 protein binds to *m-Numb* mRNA, which encodes an intracellular Notch antagonist. By inhibiting the translation of *m-Numb* mRNA, Musashi-1 augments Notch signaling and the self-renewal of NSCs. (Reproduced with the permission of Okano et al. 2002)

14.6   Blockade of Interleukin-6 Signaling in Spinal Cord Injury

It has been reported that after spinal cord injury (SCI), the intrinsic neural stem cells do not differentiate into neurons but into astrocytes, resulting in the formation of glial scars. The microenvironment of the injured spinal cord, especially the presence of inflammatory cytokines that dramatically increase in levels at the site of injury, is considered to be an important cause of inhibition of neurogenesis following SCI. Based on reports that the expression of IL-6 and the IL-6 receptor is sharply increased in the acute stages after spinal cord injury, it is considered that IL-6 may serve as a factor strongly inducing the differentiation of neural stem cells into astrocytes. In fact, the IL-6 cytokine superfamily has been demonstrated to induce NSCs to undergo astrocytic differentiation through the JAK/STAT pathway. In our recent study, we examined the effect of rat monoclonal antibody to the IL-6 receptor monoclonal antibody (MR16-1) in the treatment of acute SCI in a mouse model (Okada et al. 2004). Before conducting the in vivo experiments, we examined the effects of MR16-1 in the differentiation assay of neurosphere-derived cells and found that MR16-1 inhibited the astrocytic differentiation of NSCs in vitro (Fig. 5). For the experiment, immediately after inducing a contusion injury at the level of Th9 in mice, we administered MR-16-1 by intraperitoneal injection, which resulted in significant inhibition of activation of the JAK/STAT3 pathway. Furthermore, in the MR16-1-treated animals, there was significant reduction of astrogliosis, inflammatory cell infiltration, and connective tissue scar formation in the injured spinal cord. In addition, we observed long-lasting and significant recovery in mice treated with MR-16-1, in terms of the BBB score, as compared with control mice. These findings suggest that neutralization of the IL-6 signaling pathway in the acute phase of SCI may represent an attractive strategy for the treatment of SCI. We are currently investigating the effect of MR-16-1 treatment on the activation of endogenous NSCs and neurogenesis, based on previous findings indicating the mutually exclusive mechanisms of neuronal and astroglial differentiation of NSCs. Enhanced expression of IL-6 has also been shown to be involved in many diseases, such as rheumatoid arthritis, Crohn's disease, and Castleman's disease. A human-type monoclonal antibody directed against the IL-6 receptor has been developed jointly by Osaka University and Chugai Pharmaceutical Co., Ltd., for use in the treatment of rheumatoid arthritis. This antibody has exhibited excellent therapeutic efficacy in clinical trials (Choy et al. 2002). This human-type monoclonal antibody (MRA; Atlizumab) has been demonstrated to have potent inhibitory activity against IL-6 signals. Its safety profile and pharmacokinetic characteristics, such as its metabolism, distribution, and tolerance, have been evaluated in depth. A phase III clinical trial of this antibody is to be completed soon, and the antibody is expected to be introduced clinically in the near future; that is, the antibody is very close to the stage of clinical application. We hope that clinical trials of this antibody can also be started soon for patients with spinal cord injury.

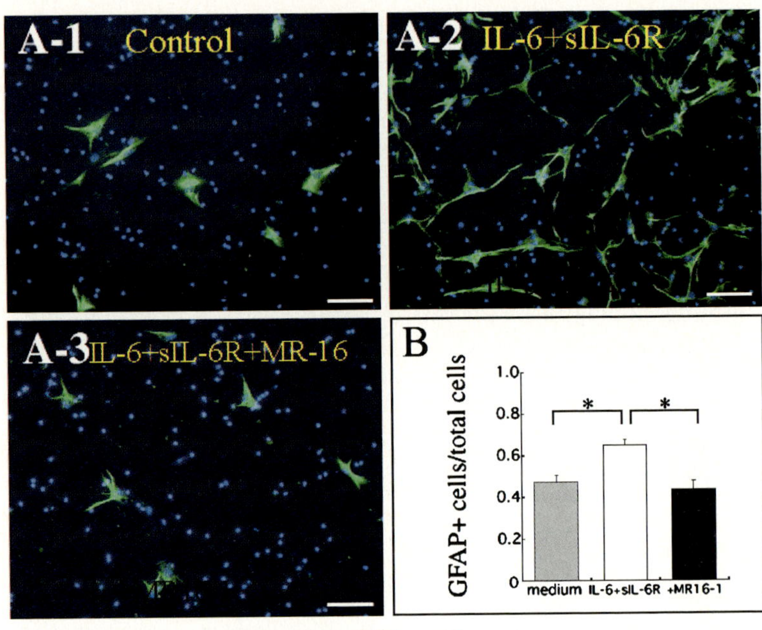

GFAP Hoechst

Fig. 5. A,B The blocking antibody of mouse IL-6 receptor (MR-16-1) inhibits the astrocytic differentiation of neurosphere cells in vitro. **A** Dissociates of neurosphere cells were cultured for 3 days with medium alone (A-1), IL-6 and soluble IL-6 (A-2), and IL-6 and soluble IL-6 plus MR16-1 (A-3) and then subjected to immunofluorescence staining for GFAP (*green*) and Hoechst 33258 (*blue*). **B** The number of GFAP-positive cells in each condition is shown. These results show that astrocytic differentiation of neurosphere cells is induced by enhancing IL-6 signaling, which is blocked by MR16-1 treatment (Reproduced with the permission of from the *Journal of Neuroscience Research*)

Acknowledgements. We would like to thank Drs. Shin-ichi Sakakibara, Takao Imai, Seiji Okada, and Masaya Nakamura for their contributions to the original research and the members of my laboratory for their valuable discussions. This work was supported by grants from the Japanese Ministry of Education, Culture, Sports, Science and Technology of Japan to HO, a grant from Terumo Foundation Life Science Foundation to HO, and a grant from the 21st Century COE program of the Japanese Ministry of Education, Culture, Sports, Science and Technology Ministry to Keio University.

References

Arvidsson A, Collin T, Kirik D, Kokaia Z, Lindvall O (2002) Neuronal replacement from endogenous precursors in the adult brain after stroke. Nat Med 8:963–970

Choy EH, Isenberg DA, Garrood T, Farrow S, Ioannou Y, Bird H, Cheung N, Williams B, Hazleman B, Price R, Yoshizaki K, Nishimoto N, Kishimoto T, Panayi GS (2002) Therapeutic benefit of blocking interleukin-6 activity with an anti-interleukin-6 receptor monoclonal antibody in rheumatoid arthritis: a randomized, double-blind, placebo-controlled, dose-escalation trial. Arthritis Rheum 46:3143–3150

Goodman CS, Doe C (1993) Embryonic development of the Drosophila central nervous system. In: Bate M, Martinez Arias A (eds) The development of *Drosophila melanogaster*. Cold Spring Harbor Laboratory Press, Cold Spring Harbor, pp 1131–1206

Imai T, Tokunaga A, Yoshida T, Hashimoto M, Weinmaster G, Mikoshiba K, Nakafuku M, Okano H (2001) The neural RNA-binding protein Musashi1 translationally regulates the m-numb gene expression by interacting with its mRNA. Mol Cell Biol 21:3888–3900

Iwanami A, Kakneko S, Nakamura M, Kanemura Y, Mori H, Kobayashi S, Yamasaki M, Momoshima S, Ishii H, Ando K, Tanioka Y, Tamaoki N, Nomura T,Toyama Y, Okano H (2005a) Transplantation of human neural stem/progenitor cells promotes functional recovery after spinal cord injury in common marmoset. J Neurosci Res 80:182–190

Iwanami A, Yamane J, Katoh H, Nakamura M, Momomoshima S, Ishii H, Tanioka Y, Tamaoki N, Nomura T, Toyama Y, Okano H (2005b) Establishment of graded spinal cord injury model in a non-human primate: the common marmoset. J Neurosci Res 80:172–181

Kaneko Y, Sakakibara S, Imai T, Suzuki A, Nakamura Y, Sawamoto K, Ogawa Y, Toyama Y, Miyata T, Okano H (2000) Musashi 1: an evolutionally conserved marker for CNS progenitor cells including neural stem cells. Dev Neurosci 22:138–152

Kawaguchi A, Miyata T, Sawamoto K, Takashita N, Murayama A, Akamatsu W, Ogawa M, Okabe M, Tano Y, Goldman SA, Okano H (2001) Nestin-EGFP transgenic mice: visualization of the self-renewal and multipotency of CNS stem cells. Mol Cell Neurosci 17:259–273

Kempermann G, Gage FH (1999) New nerve cells for the adult brain. Sci Am 280:48–53

Keyoung HM, Roy NS, Benraiss A, Louissaint A Jr, Suzuki A, Hashimoto M, Rashbaum WK, Okano H, Goldman SA (2001) Prospective identification, selection and extraction of two distinct pools of neural stem cells from fetal human brain. Nat Biotechnol 19:843–850

Kohyama J, Tokunaga A, Fujita Y, Miyoshi H, Nagai T, Miyawaki A, Nakao K, Mastuzaki Y, Okano H (2005) Visualization of spatio-temporal activation of Notch signaling: live monitoring and significance in neural development. Dev Biol 286:311–325

Lindvall O, Kokaia Z, Martinez-Serrano A (2004) Stem cell therapy for human neurodegenerative disorders-how to make it work. Nat Med 10 [Suppl]:S42–S50

Matsuzaki Y, Kinjo K, Mulligan RC, Okano H (2004) Unexpectedly efficient homing capacity of purified murine hematopoietic stem cells. Immunity 20:87–93

Nakamura M, Houghtling RA, MacArthur L, Bayer BM, Bregman BS (2003) Differences in cytokine gene expression profile between acute and secondary injury in adult rat spinal cord. Exp Neurol 184:313–325

Nakamura Y, Sakakibara S, Miyata T, Ogawa M, Shimazaki T, Weiss S, Kageyama R, Okano H (2000) The bHLH gene Hes1 as a repressor of neuronal commitment of the CNS stem cells. J Neurosci 20:283–293

Ogawa Y, Sawamoto K, Miyata T, Miyao S, Watanabe M, Toyama Y, Nakamura M, Bregman BS, Koike M, Uchiyama Y, Toyama Y, Okano H (2002) Transplantation of in vitro expanded fetal neural progenitor cells results in neurogenesis and functional recovery after spinal cord contusion injury in rats. J Neurosci Res 69:925–933

Okada S, Nakamura M, Mikami Y, Shimazaki T, Mihara M, Ohsugi Y, Iwamoto Y, Yoshizaki K, Kishimoto T, Toyama Y, Okano H (2004) Blockade of interleukin-6 receptor suppresses reactive astrogliosis and ameliorates functional recovery in experimental spinal cord injury. J Neurosci Res 76:265–276

Okano H (2002a) Neural stem cells: progression of basic research and perspective for clinical application. Keio J Med 51:115–128

Okano H (2002b) The stem cell biology of the central nervous system. J Neurosci Res 69:698–707

Okano H (2003) Making and repairing the mammalian brain: introduction. Semin Cell Dev Biol 14:159

Okano H, Imai T, Okabe M (2002) Musashi: a translational regulator of cell fates. J Cell Sci 115:1355–1359

Okano H, Ogawa Y, Nakamura M, Kaneko S, Iwanami A, Toyama A (2003) Transplantation of neural stem cells into the spinal cord after injury. Semin Cell Dev Biol 14:191–198

Okano H, Kawahara H, Toriya M, Nakao K, Shibata S, Takao I (2005) Function of RNA binding protein Musashi-1 in stem cells. Exp Cell Res 306:349–356

Osawa M, Hanada K, Hamada H, Nakauchi H (1996) Long-term lymphohematopoietic reconstitution by a single CD34-low/negative hematopoietic stem cell. Science 273:242–245

Pincus D, Keyoung H, Restelli C, Sakakibara S, Okano H, Goodman R, Fraser R, Edgar M, Nedergaard M, Goldman SA (1998) FGF2/BDNF-associated maturation of new neurons generated from adult human subependymal cells. Ann Neurol 43:576–585

Ramón y Cajal S (1928) Degeneration and regeneration of the nervous system (Translated by RM Day from the 1913 Spanish edition). Oxford: Oxford University Press

Reynolds BA, Weiss S (1992) Generation of neurons and astrocytes from isolated cells of the adult mammalian central nervous system. Science 255:1707–1710

Rietze RL, Valcanis H, Brooker GF, Thomas T, Voss AK, Bartlett PF (2001) Purification of a pluripotent neural stem cell from the adult mouse brain. Nature 412:736–739

Roy NS, Wang S, Jiang L, Kang J, Restelli C, Fraser RAR, Couldwell WT, Kawaguchi A, Okano H, Nedergaard M, Goldman SA (2000) In vitro neurogenesis by neural progenitor cells isolated from adult human hippocampus. Nat Med 6:271–278

Sakakibara S, Okano H (1997) Expression of neural RNA-binding proteins in the post-natal CNS: implication of their roles in neural and glial cell development. J Neurosci 17:8300–8312

Sakakibara S, Imai T, Aruga J, Nakajima K, Yasutomi D, Nagata T, Kurihara Y, Uesugi S, Miyata T, Ogawa M, Mikoshiba K, Okano H (1996) Mouse-Musashi-1, a neural RNA-binding protein highly enriched in the mammalian CNS stem cell. Dev Biol 176:230–242

Sakakibara S, Nakamura Y, Yoshida T, Shibata S, Koike M, Takano H, Ueda S, Uchiyama Y, Noda T, Okano H (2002) RNA-binding protein Musashi family: roles for CNS stem cells and a subpopulation of ependymal cells revealed by targeted disruption and antisense ablation. Proc Natl Acad Sci U S A 99:15194–15199

Sawamoto K, Nakao N, Kakishita K, Ogawa Y, Toyama Y, Yamamoto A, Yamaguchi M, Mori K, Goldman SA, Itakura T, Okano H (2001) Generation of dopaminergic neurons in the adult brain from mesencephalic precursor cells labeled with a nestin-GFP transgene. J Neurosci 21:3895–3903

Temple S, Alvarez-Buylla A (1999) Stem cells in the adult mammalian central nervous system. Curr Opin Neurobiol 9:135–141

Tonchev AB, Yamashima T, Sawamoto K, Okano H (2005) Enhanced proliferation of progenitor cells in the subventricular zone and limited neuronal production in the striatum and neocortex of adult macaque monkeys after global cerebral ischemia. J Neurosci Res 81:776–788
Uchida N, Buck DW, He D, Reitsma MJ, Masek M, Phan TV, Tsukamoto AS, Gage FH, Weissman IL (2000) Direct isolation of human central nervous system stem cells. Proc Natl Acad Sci U S A 97:14720–14725
Yagita Y, Kitagawa K, Sasaki T, Miyata T, Okano H, Hori M, Matsumoto M (2000) Differential expression of Musashi1 and nestin in the adult rat hippocampus after ischemia. J Neurosci Res 69:750–756

15 Stem Cell Therapy for Parkinson's Disease

J. Takahashi

15.1   Introduction 230
15.2   Region Specificity of Human Neural Precursor Cells 231
15.3   Dopaminergic Differentiation from Embryonic Stem Cells 233
15.4   Transplantation of Dopaminergic Neurons Derived
        from Embryonic Stem Cells 234
15.5   Conclusion 239
References 240

Abstract. Transplantation of fetal dopaminergic (DA) neurons can produce symptomatic relief for patients with Parkinson's disease, but the technical and ethical difficulties have limited the application of this therapy. Neural precursor cells and embryonic stem cells (ESCs) are expected to be candidates of potential donor cells for transplantation. Human neural precursor cells obtained from the midbrain give rise to TH-positive neurons. The growth of the cells, however, is slow and the differentiation rate of DA neurons is still low for clinical application. Monkey ESCs give rise to midbrain DA neurons, and the transplanted ESC-derived neurospheres function as DA neurons, attenuating the neurological symptoms of the monkey Parkinson's disease model. These results suggest the possibility of using stem cells for the treatment of Parkinson's disease, but problems such as the low survival rate in vivo and tumor formation must be solved.

15.1 Introduction

Parkinson's disease (PD) is a neurodegenerative disorder characterized by a loss of midbrain dopaminergic (DA) neurons and a subsequent reduction in striatal DA (Olanow and Tatton 2000). While initial pharmacological treatment with L-dihydroxyphenylalanine (L-DOPA) has proven effective, the efficacy of this treatment progressively declines and motor complications such as dyskinesia may also develop. As a result, alternative approaches such as deep brain stimulation and fetal DA neuron transplantation (Bergman and Deuschl 2002; Dostrovsky et al. 2002; Miyasaki et al. 2002) are necessary for treatment of PD. Studies using animal models and clinical investigations have shown that transplantation of fetal DA neurons can produce symptomatic relief (Freed et al. 2001; Hagell et al. 2001; Lindvall et al. 1994; Olanow et al. 2003; Widner et al. 1992). However, the technical and ethical difficulties involved in obtaining sufficient and appropriate donor fetal brain tissue have limited the application of this therapy.

In 2001, the results of a double-blinded, sham surgery-controlled trial of the transplantation of human embryonic DA neurons into patients with PD were reported (Freed et al. 2001). Among patients 60 years old or younger, standardized evaluations of PD revealed a significant improvement in the transplantation group. However, cells from at least four aborted human embryos were required per patient in order to produce symptomatic recovery, making this procedure unfavorable for treating large numbers of patients. In an attempt to increase the supply of DA neurons, rat fetal ventral mesencephalic cells, which include DA neuronal precursors, were expanded in vitro and then transplanted into rats with PD symptoms. Although symptomatic recovery was observed, the cell survival rate was as low as 3%–5% in vivo, offsetting any advantage gained by grafting in vitro-expanded cells rather than fresh (unexpanded) fetal ventral mesencephalon (Studer et al. 1998). These results provided a strong impetus to seek alternative cell sources that could be expanded in vitro.

15.2 Region Specificity of Human Neural Precursor Cells

Human neural precursor cells (NPCs), namely, neural stem cells and neural progenitor cells, have also been isolated from fetal and adult brain (Carpenter et al. 1999; Ostenfeld et al. 2000; Palmer et al. 2001; Svendsen et al. 1997; Vescovi et al. 1999). These cells give rise to neurons and glial cells in vitro and survive to differentiate into neurons in the rat brain (Fricker et al. 1999). These findings support the idea that the NPC is a good candidate for cell transplantation therapy against various neurological diseases.

Recent evidence suggests that murine and human NPCs have different potentials to proliferate and differentiate by their region of origin (Hitoshi et al. 2002; Jain et al. 2003; Ostenfeld et al. 2002). To clarify the suggestions, we examined the potential of NPCs derived from human embryo (Horiguchi et al. 2004; Fig. 1). To determine the regional differences of these precursor cells, we divided the brain of a 9-week-old human embryo into four parts: telencephalon, diencephalon, mesencephalon, and rhombencephalon. In this study, two region-specific differences in NPCs were revealed: (1) The precursor cells from the rostral part of the brain tended to proliferate faster than those from the caudal part, and (2) The precursor cells from the diencephalon and mesencephalon gave rise to more TH-positive neurons than those from the telencephalon and rhombencephalon. Furthermore, the former TH-positive cells were large, multipolar, and GABA-negative, which suggested that these cells were midbrain DA neurons. In contrast, the latter were small, bipolar, and GABA-positive. Virtually all the TH-positive periglomerular neurons in the rat olfactory bulb are also immunoreactive for GABA (Gall et al. 1987). There are newborn neurons immunoreactive for both TH and GABA in the olfactory bulb of newborn rats at postnatal day 2 (Betarbet et al. 1996). Consequently, it is possible that TH-positive cells from telencephalic precursors were such small neurons of the olfactory bulb. In agreement with our results, recent studies demonstrated that human mesencephalic precursor cells gave rise to significantly more TH-positive neurons than telencephalic precursors in response to cytokine/GDNF mixture (Storch et al. 2001) or after transplantation into the rat brain (Sanchez-Pernaute et al. 2001). These results demonstrated that multipotent NPCs can be expanded from the human

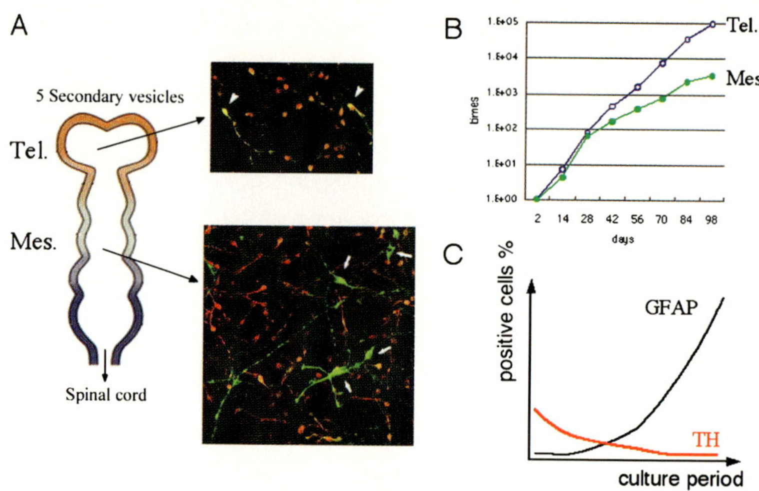

Fig. 1. A–C Neural precursor cells (NPCs) derived from human embryo **A** Differentiated NPCs for TH (*green*) and GABA (*red*) were double-labeling immunofluorescence stained. Most of the TH-positive cells in the mesencephalon cultures were large, multipolar, and GABA negative (*arrows*). In contrast, TH-positive cells in the telencephalon cultures were small, bipolar, and GABA-positive (*arrowheads*). **B** Parallel cultures of NPCs derived from telencephalon, and mesencephalon were maintained in hFGF-2, hEGF, and hLIF. The ratios of total cell numbers to those on day 2 in each culture were plotted on a semilogarithmic graph. **C** Neurospheres derived from human embryonic brain were cultured for up to 150 days and then made to differentiate for 14 days. The percentages of TH- and GFAP-positive cells per total cells on each culture period are shown. (Reproduced in part, with permission, from Horiguchi et al. 2004)

embryonic brain and that the cells give rise to TH-positive neurons. But because of regional specificity, NPCs derived from the mesencephalon are required to obtain midbrain DA neurons. Furthermore, the percentage of DA neurons drops to 0.2% of total cells while that of astrocytes increases up to 50% after a long-term culture. One of the advantages of using NPCs for cell transplantation is that they can be expanded almost unlimitedly. But these results suggest that human NPCs expanded for a long time period might mainly originate from telencephalon and give rise to few DA neurons and many astrocytes. For an effective cell transplantation therapy for Parkinson's disease, selective sorting and/or selective proliferation of dopaminergic progenitor cells will be necessary.

15.3 Dopaminergic Differentiation from Embryonic Stem Cells

In 2000, two groups reported methods for generating DA neurons from mouse ESCs, thus opening the way to unlimited in vitro production of these neurons. Lee et al. (2000) introduced a five-step method in which undifferentiated ESCs (step 1) were differentiated into embryoid bodies on nonadherent culture plates for 4 days in the presence of serum (step 2). The embryoid bodies were then plated on adhesive tissue culture plates for 24 h in the presence of serum and transferred to serum-free medium for 4–6 days to select for neural precursor cells (step 3). These cells were dissociated and grown for 6 days on an adhesive substrate in the presence of FGF8, Sonic hedgehog (Shh), and bFGF (step 4). Finally, bFGF was removed and ascorbic acid was added for an additional 6–15 days (step 5). The resulting neurons expressed DA neuronal markers Nurr1 and tyrosine hydroxylase (TH), the rate-limiting enzyme in dopamine synthesis. Furthermore, these neurons released dopamine in response to depolarization with potassium. Under optimal culture conditions, 71.9% of the ESCs assumed a neuronal morphology, 33.9% of which were dopaminergic. Since cells were grown in chemically defined media, with serum added only at step 1 and step 2, this technique may be applicable to clinical therapy in the near future. However, despite extensive work so far, midbrain DA neurons have not been effectively induced from primate or human ES cells using the five-step method (Kuo et al. 2003).

Independently, Kawasaki et al. (2000) reported another method for generating DA neurons from ES cells. In contrast to the five-step method, this method does not require growth in serum, the formation of embryoid bodies, or the selection of neural precursor cells. Instead, the authors demonstrated that the bone marrow-derived stromal cell line PA6 is a potent inducer of neural differentiation from ESCs. They named this activity stromal cell-derived inducing activity (SDIA). After coculture with PA6 cells for 14 days, 52% of mouse ESCs differentiated into neu-

rons, 30% of which were dopaminergic. Intriguingly, the SDIA method
is also applicable to primate ES cells, although the efficacy of neural
and dopaminergic differentiation is about half of that observed with
mouse ESCs (Kawasaki et al. 2002). Recently, Perrier et al. (2004)
succeeded in inducing DA neuron differentiation from human ESCs by
modifying the SDIA method, which were also confirmed by other inves-
tigators (Buytaert-Hoefen et al. 2004; Zeng et al. 2004). Thus, the SDIA
method is a promising step toward the clinical application of human
ESC therapy. However, because PA6 feeder cells are mouse-derived, the
molecular mechanisms of SDIA must be elucidated so that the signaling
molecules responsible for inducing DA neuron differentiation can be
directly applied to ESCs.

15.4 Transplantation of Dopaminergic Neurons Derived from Embryonic Stem Cells

Once DA neurons are efficiently generated from ESCs, these cells must
be shown to function in the brain to relieve the symptoms of Parkinson's disease. Björklund et al. (2002) reported that low concentration,
single-cell suspensions of embryoid bodies derived from mouse ESCs
proliferate and spontaneously differentiate into DA neurons when transplanted into the 6-hydroxydopamine (OHDA)-lesioned rat striatum and
can rescue motor asymmetry defects in Parkinson's disease models.
However, they also observed overproliferation of the grafted cells, resulting in teratoma-like tumor formation in five (20%) of the 25 recipient
animals. Because mesodermal, endodermal, and even undifferentiated
ESCs are contained in embryoid bodies and can cause tumor formation,
methods to exclude these non-neural cells must be established in order
to avoid tumorigenesis as a side effect of cell transplantation.

In addition, Kim et al. (2002) observed a significant behavioral improvement in mouse PD models following transplantation of DA neurons
generated by the five-step method from ESCs overexpressing Nurr1,
a transcription factor that plays an important role in the differentiation
of midbrain precursors into DA neurons. They observed that about 80%
of the ESC-derived neurons were TH-positive, twice the percentage observed using the conventional five-step method, resulting in 56% of total

cells differentiating into DA neurons. When transplanted into the striatum of 6-OHDA-lesioned rats, these DA neurons showed electrophysiological activity in vivo and were able to rescue the rotational asymmetry defects associated with ipsilateral degeneration of DA neurons. A similar improvement was reported from transplantation of ESC derivatives differentiated by a modified SDIA protocol (Barberi et al. 2003).

As mentioned above, the SDIA method can efficiently induce DA neurons from primate ES cells. Applying these cells to a monkey model of PD, we demonstrated that DA neurons induced from monkey ESCs reduce parkinsonian symptoms when transplanted into animals (Takagi et al. 2005; Fig. 2). After coculture of monkey ESCs with PA6 cell

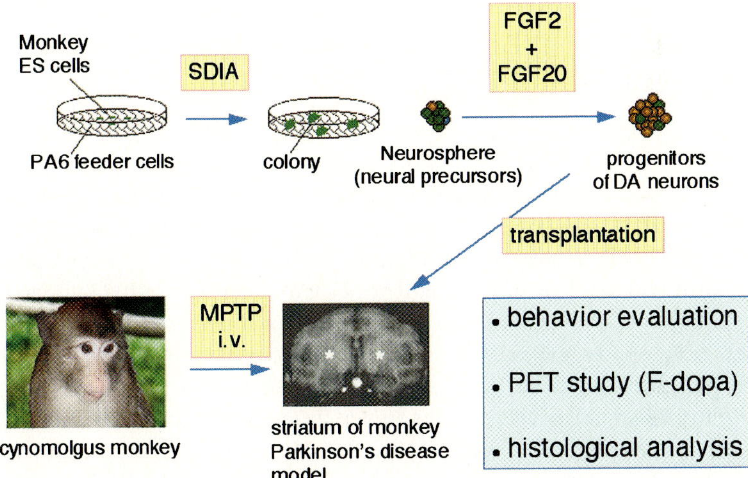

Fig. 2. A schema of the monkey Parkinson's disease model. Neurospheres composed of neural progenitors, which were capable of producing large numbers of DA neurons, were generated from monkey embryonic stem cells (ESCs). FGF20 acted synergistically with FGF2 to increase the number of dopaminergic (DA) neurons in ESC-derived neurospheres, suggesting that FGF2 and FGF20 increase the number of DA neuron progenitors in the sphere. The effect of transplantation of DA neurons generated from monkey ESCs into 1-methyl-4-phenyl-1,2,3,6-tetrahydropyridine (MPTP)-treated monkeys, a primate model for Parkinson's disease, was also analyzed

lines for 14 days, the authors detached the ESC derivatives and cultured them as floating aggregates in the presence of FGF-2 and FGF-20 for additional 7 days. FGF20 is a member of the FGF family that is expressed exclusively in the substantia nigra of the midbrain and is reported to have a supporting effect on midbrain DA neurons (Ohmachi et al. 2000). This treatment caused an approximately fourfold increase in TH-positive neurons compared to the conventional SDIA method. A primate model of Parkinson's disease has already been established by repetitive injection of the DA-specific neurotoxin 1-methyl-4-phenyl-1,2,3,6-tetrahydropyridine (MPTP), which results in many symptoms characteristic of human Parkinson's disease such as tremor, rigidity, akinesia, and gait disturbance (Burns et al. 1983). When transplanted into the striatum of MPTP-treated monkeys, monkey ESCs treated by SDIA and FGF-2/FGF-20 survived and functioned as DA neurons, resulting in behavioral recovery and a decrease in parkinsonian symptoms. Increased uptake of ^{18}F-fluorodopa observed by positron emission tomography (PET) confirmed that the grafted cells functioned as dopaminergic neurons (Fig. 3).

Although DA neurons induced from both mouse and primate ESCs are able to treat parkinsonian symptoms in animal models, problems remain. First, ESCs differentiate heterogeneously and do not behave synchronously, resulting in contamination of ESC-derived grafts with undifferentiated ESCs (Kawasaki et al. 2000; Ying et al. 2003). Transplantation of undifferentiated ESCs or embryoid bodies into adult brain has been shown to result in the formation of teratomas or teratocarcinomas (Björklund et al. 2002; Erdö et al. 2003). In most reports so far, teratoma formation was not reported after transplantation of differentiated neural cell grafts derived from mouse and human ESCs (Ben-Hur et al. 2004; Kim et al. 2002; Reubinoff et al. 2001; Zhang et al. 2001). These studies, however, were xenograft transplantations (for example, mouse cells into a rat host), suggesting that the lack of tumor formation in these experiments might be caused by severe immunorejection due to transplantation of cells from different species. Indeed, allograft experiments in which mouse ESC-derived neurons were transplanted into a mouse host resulted in tumor formation or tumor-like cell growth, even though the cells had been differentiated for 14 days or more in vitro, possibly due to a small fraction of undifferentiated ESCs in the graft

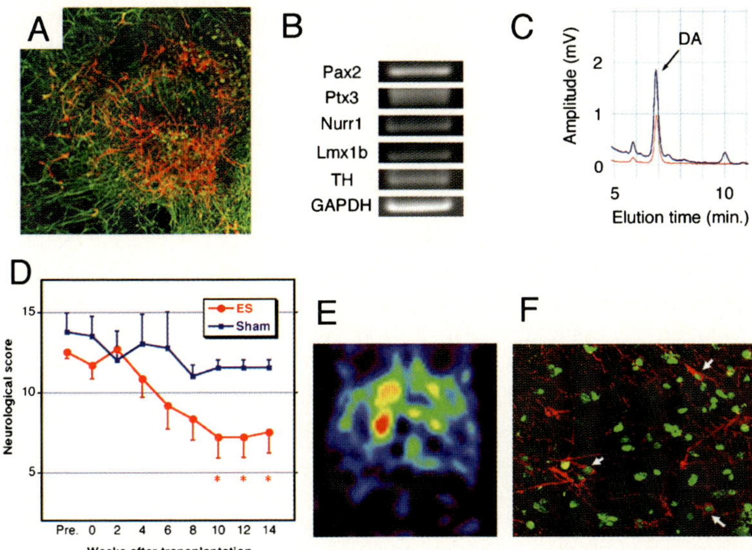

Fig. 3. A–F Results from transplantation of cynomolgus monkey ESC-derived grafts into MPTP-treated monkey models of Parkinson's disease A Numerous TH-positive DA neurons (*red*; *green* = Tuj1) were generated from cynomolgus monkey ESCs by a stromal cell-derived inducing activity (SDIA) protocol with minor modifications. B Expression of Pax2, Ptx3, Nurr1, and Lmx1b suggests that grafted cells have midbrain DA neuronal character. C Dopamine release from differentiated cells in response to high K^+ stimuli, assayed by HPLC (high performance liquid chromatography). D Behavioral scores showed a significant improvement in the transplantation group ($n = 6$), compared to the sham-operated group ($n = 4$). E In the PET study at 14 weeks, an axial image of the monkey brain receiving grafts into the bilateral putamen showed increased uptake of ^{18}F-fluorodopa on both sides (R>L), suggesting that the transplanted cells are functionally active. F Immunohistochemistry revealed that grafted cells (BrdU-positive cells incorporated prior to transplantation: *green*) survived in the brain and differentiated into DA neurons (TH: *red*). *Arrows* indicate some of the BrdU/TH-positive cells, demonstrating that these cells are ESC-derived DA neurons. (Assembled, with permission, from the figures in Takagi et al. 2005)

(Barberi et al. 2003; Erdö et al. 2003; Nishimura et al. 2003). Therefore, it is essential for undifferentiated ESCs to be eliminated from the graft in order for transplantation to be feasible. Recently, Ying et al. (2003) reported that ESCs can be differentiated into neural precursors by an adherent monolayer culture, without the formation of embryoid bodies or growth on feeder cells. In this report, the authors used fluorescence-activated cell sorting (FACS) to purify neural precursors differentiated from a Sox1-GFP knock-in reporter ESC line. Sox1 is the earliest known specific marker of neuroectoderm in the mouse embryo (Wood et al. 1999). Another group has succeeded in purifying ESC-derived neural precursors by inducing selective apoptosis of undifferentiated ESCs using novel ceramide analogues (Bieberich et al. 2004). Thus, both optical and pharmacological technologies to eliminate undifferentiated ESCs are being developed to support the possible application of human ESCs to transplantation therapy, although further investigation will be required to completely prevent tumor formation as a side effect.

Another problem to be addressed is the low survival rate of the grafted cells in the brain. In animal experiments, the survival rate of transplanted DA neurons is extremely low (0.8%–5%) (Studer et al. 1998; Takagi et al. 2005). This is probably caused by the mechanical damage during cell preparation and injection, lack of trophic factors, and even acute inflammation. Since apoptosis appears to contribute greatly to cell death following transplantation, inhibition of apoptosis should improve the cell survival rate. Indeed, Schierle et al. reported that treatment of a rat nigral cell suspension with the caspase inhibitor acetyl-tyrosinyl-valyl-alanyl-aspartyl-chloro-methylketone (Ac-YVAD-cmk) prior to transplantation effectively blocked DNA fragmentation (apoptosis), resulting in both a 16-fold decrease in cell death at the acute stage and a threefold increase in surviving DA neurons 6 weeks after transplantation (Schierle et al. 1999).

Controversy remains on whether immunosuppression is indispensable, because allografts of fetal DA neurons have not been rejected without immunosuppression in some reports (Freed et al. 2001; Studer et al. 1998). However, rejection by the host immune system (if any) may also prevent effective transplantation. Persistent administration of immunosuppressive drugs will likely compromise the quality of patients' lives, as suggested by cases of organ transplantation. To bypass this problem, mouse ESC lines have been generated from somatic cells by nuclear transfer (nt ESCs), enabling us to achieve virtually autologous transplantation without immunosuppressive drugs (Wakayama et al. 2001). Efficient differentiation into DA neurons from mouse nt ES cells has already been confirmed (Barberi et al. 2003). However, it is not currently known whether nt ESCs are safe and whether DA neurons differentiated from nt ESCs have the same lifespan as those from conventional ESCs (the lifespan of the cell is likely programmed in the nucleus).

15.5   Conclusion

In summary, ESC transplantation therapy for Parkinson's disease should be feasible if methods are found to establish and expand ESCs under defined conditions. So far the variety of differentiation methods to induce DA neurons from ESCs have been tried, and these are still evolving. The methods to prevent tumor formation by purifying DA neurons, to overcome immunological obstacles by therapeutic cloning or other means, are also being developed.

Until now, a great deal of work has focused on donor cells and graft methods; however, future research should next focus on host factors, since creating a permissive environment for the graft to integrate into the host brain is likely of equal importance to supplying the ideal graft. Some evidence suggests that aging (Conboy et al. 2003; Kuhn et al. 1996; Morrison et al. 1996), inflammation (Monje et al. 2003), and molecular signals inducing the graft into glial cell fate (Lim et al. 2000) are involved in both reduced tissue regenerative potential and inefficient graft integration. Research investigating the molecular mechanisms of these phenomena will provide new insights that will contribute to the success of cell transplantation therapies (Conboy et al. 2005; Iakova et al. 2003; Lim et al. 2000; Monje et al. 2003).

As a result of great efforts over the last decade, stem cell research may now be poised to bring what was heretofore considered impossible into the realm of reality. We hope that continued work in the future will make cell transplantation therapy for Parkinson's disease as commonplace a procedure as bone marrow transplantation, a currently successful example of cell transplantation therapy.

Acknowledgements. The author wishes to thank the Editors of Expert Opinion on Biological Therapy for generous permission to copy a part of the text passages from the author's article to be published in the journal. The author's work was supported by the following grants: Grants-in-Aid for Scientific Research from the JSPS; Special Coordination Funds for Promoting Science and Technology, Grants in Kobe Cluster, and Establishment of International COE for Integration of Transplantation Therapy and Regenerative Medicine from the MEXT; Health Sciences Research Grants in Research on Human Genome, Tissue Engineering and Food Biotechnology from the MHLW; and Grants in Organization for Pharmaceutical Safety and Research.

References

Barberi T, Klivenyi P, Calingasan NY, Lee H, Kawamata H, Loonam K, Perrier AL, Bruses J, Rubio ME, Topf N, Tabar V, Harrison NL, Beal MF, Moore MAS, Studer L (2003) Neural subtype specification of fertilization and nuclear transfer embryonic stem cells and application in parkinsonian mice. Nat Biotechnol 10:1200–1207

Ben-Hur T, Idelson M, Khaner H, Pera M, Reinhartz E, Itzik A, Reubinoff BE (2004) Transplantation of human embryonic stem cell-derived neural progenitors improves behavioral deficit in parkinsonian rats. Stem Cells 22:1246–1255

Bergman H, Deuschl G (2002) Pathophysiology of Parkinson' disease: from clinical neurology to basic neuroscience and back. Mov Disord 17 Suppl. 3:S28–S40

Betarbet R, Zigova T, Bakay RA, Luskin MB (1996) Dopaminergic and GABAergic interneurons of the olfactory bulb are derived from the neonatal subventricular zone. Int J Dev Neurosci 14:921–930

Bieberich E, Silva J, Wang G, Krishnamurthy K, Condie BG (2004) Selective apoptosis of pluripotent mouse and human stem cells by novel ceramide analogues prevents teratoma formation and enriches for neural precursors in ES cell-derived neural transplants. J Cell Biol 167:723–734

Björklund LM, Sánchez-Pernaute R, Chung S, Andersson T, Che IYC, McNaught KSTP, Brownell A-L, Jenkins BG, Wahlestedt C, Kim K-S, Isacson O (2002) Embryonic stem cells develop into functional dopaminergic neurons after transplantation in a Parkinson rat model. Proc Natl Acad Sci U S A 99:2344–2349

Burns RS, Chiueh CC, Markey SP, Ebert MH, Jacobowitz DM, Kopin IJ (1983) A primate model of parkinsonism: selective destruction of dopaminergic neurons in the pars compacta of the substantia nigra by N-methyl-4-phenyl-1,2,3,6-tetrahydropyridine. Proc Natl Acad Sci U S A 80:4546–4550

Buytaert-Hoefen KA, Alvarez E, Freed CR (2004) Generation of tyrosine hydroxylase positive neurons from human embryonic stem cells after coculture with cellular substrates and exposure to GDNF. Stem Cells 22:669–674

Carpenter MK, Cui X, Hu Z-Y, Jackson J, Sherman S, Seiger A, Wahlberg LU (1999) In vitro expansion of a multipotent population of human neural progenitor cells. Exp Neurol 158:265–278

Conboy IM, Conboy MJ, Smythe GM, Rando TA (2003) Notch-mediated restoration of regenerative potential to aged muscle. Science 302:1575–1577

Conboy IM, Conboy MJ, Wagers AJ, Girma ER, Weissman IL, Rando TA (2005) Rejuvenation of aged progenitor cells by exposure to a young systemic environment. Nature 433:760–764

Dostrovsky JO, Hutchinson WD, Lozano AM (2002) The globus pallidus, deep brain stimulation, and Parkinson's disease. Neuroscientist 8:284–290.

Erdö F, Bührle C, Blunk J, Hoehn M, Xia Y, Fleischmann B, Föcking M, Küstermann E, Kolossov E, Hescheler J, Hossmann K-A, Trapp T (2003) Host-dependent tumorigenesis of embryonic stem cell transplantation in experimental stroke. J Cereb Blood Flow Mctab 23:780–785

Freed CR, Greene PE, Breeze RE, Tsai WY, DuMouchel W, Kao R, Dillon S, Winfield H, Culver S, Trojanowski JQ, Eidelberg D, Fahn S (2001) Transplantation of embryonic dopamine neurons for severe Parkinson's disease. N Engl J Med 344:710–719

Fricker RA, Carpenter MK, Winkler C, Greco C, Gates MA, Björklund A (1999) Site-specific migration and neural differentiation of human neural progenitor cells after transplantation in the adult rat brain. J Neurosci 19:5990–6005

Gall CM, Hendry SH, Seroogy KB, Jones EG, Haycock JW (1987) Evidence for coexistence of GABA and dopamine in neurons of the rat olfactory bulb. J Comp Neurol 266:307–318

Hagell P, Brundin P (2001) Cell survival and clinical outcome following intrastriatal transplantation in Parkinson disease. J Neuropathol Exp Neurol 60:741–752

Horiguchi S, Takahashi J, Kishi Y, Morizane A, Okamoto Y, Koyanagi M, Tsuji M, Tashiro K, Honjo T, Fujii S, Hashimoto N (2004) Neural precursor cells derived from human embryonic brain retain regional specificity. J Neurosci Res 75:817–824

Iakova P, Awad SS, Timchenko NA (2003) Aging reduces proliferative capacities of liver by switching pathways of C/EBPa growth arrest. Cell 113:495–506

Kawasaki H, Mizuseki K, Nishikawa S, Kaneko S, Kuwana Y, Nakanishi S, Nishikawa S-I, Sasai Y (2000) Induction of midbrain dopaminergic neurons from ES cells by stromal cell-derived activity. Neuron 28:31–40

Kawasaki H, Suemori H, Mizuseki K, Watanabe K, Urano F, Ichinose H, Haruta M, Takahashi M, Yoshikawa K, Nishikawa S-I, Nakatsuji N, Sasai Y (2002) Generation of dopaminergic neurons and pigmented epithelia from primate ES cells by stromal cell-derived inducing activity. Proc Natl Acad Sci U S A 99:1580–1585

Kim J-H, Auerbach JM, Rodriguez-Gómez J, Velasco I, Gavin D, Lumelsky N, Lee S-H, Nguyen J, Sánchez-Pernaute R, Banliewicz K, McKay RD (2002) Dopamine neurons derived from embryonic stem cells function in an animal model of Parkinson's disease. Nature 418:50–56

Kuhn HG, Dickson-Anson H, Gage FH (1996) Neurogenesis in the dentate gyrus of the adult rat: age-related decrease of neuronal progenitor proliferation. J Neurosci 16:2027–2033

Kuo HC, Pau KY, Yeoman RR, Mitalipov SM, Okano H, Wolf DP (2003) Differentiation of monkey embryonic stem cells into neural lineages. Biol Reprod 68:1727–1735

Lee SH, Lumelsky N, Studer L, Auerbach JM, McKay RD (2000) Efficient generation of midbrain and hindbrain neurons from mouse embryonic stem cell. Nat Biotech 18:675–679

Lim DA, Tramontin AD, Trevejo JM, Herrera DG, Garcia-Verdugo JM, Alvarez-Buylla A (2000) Noggin antagonizes BMP signaling to create a niche for adult neurogenesis. Neuron 28:713–726

Lindvall O, Sawle G, Widner H, Rothwell JC, Bjorklund A, Brooks D, Brundin P, Frackowiak R, Marsden CD, Odin P (1994) Evidence for long-term survival and function of dopaminergic grafts in progressive Parkinson's disease. Ann Neurol 35:172–180

Miyasaki JM, Martin W, Suchowersky O, Weiner WJ, Lang AE (2002) Practice parameter: initiation of treatment for Parkinson's disease: an evidence-based review: report of the Quality Standards Subcommittee of the American Academy of Neurology. Neurology 58:11–17

Monje L, Toda H, Palmer TD (2003) Inflammatory blockade restores adult hippocampal neurogenesis. Science 302:1760–1765

Morrison SJ, Wandycz AM, Akashi K, Globerson A, Weissman IL (1996) The aging of hematopoietic stem cells. Nat Med 2:1011–1016

Nishimura F, Yoshikawa M, Kanda S, Nonaka M, Yokota H, Shiroi A, Nakase H, Hirabayashi H, Ouji Y, Birumachi J, Ishizaka S, Sakaki T (2003) Potential use of embryonic stem cells for the treatment of mouse parkinsonian models: improved behavior by transplantation of in vitro differentiated dopaminergic neurons from embryonic stem cells. Stem Cells 21:171–180

Ohmachi S, Watanabe Y, Mikami T, Kusu N, Ibi T, Akaike A, Itoh N (2000) FGF-20, a novel neurotrophic factor, preferentially expressed in the substantia nigra pars compacta of rat brain. Biochem Biophys Res Commun 277:355–360

Olanow CW, Tatton WG (2000) Etiology and pathogenesis of Parkinson's disease. Annu Rev Neurosci 22:123–144

Olanow CW, Goetz CG, Kordower JH, Stoessl AJ, Sossi V, Brin MF, Shannon KM, Nauert GM, Perl DP, Godbold J, Freeman TB (2003) A double blind controlled trial of bilateral fetal nigral transplantation in Parkinson's disease. Ann Neurol 54:403–414

Ostenfeld T, Caldwell MA, Prowse KR, Linskens MH, Jauniaux E, Svendsen CN (2000) Human neural precursor cells express low levels of telomerase in vitro and show diminishing cell proliferation with extensive axonal outgrowth following transplantation. Exp Neurol 164:215–226

Ostenfeld T, Joly E, Tai Y-T, Peters A, Caldwell M, Jauniaux E, Svendsen CN (2002) Regional specification of rodent and human neurospheres. Dev Brain Res 134:43–55

Palmer TD, Schwartz PH, Taupin P, Kaspar B, Stein SA, Gage FH (2001) Progenitor cells from human brain after death. Nature 411:42–43

Perrier AL, Tabar V, Barberi T, Rubio ME, Bruses J, Topf N, Harrison NL, Studer L (2004) Derivation of midbrain dopamine neurons from human embryonic stem cells. Proc Natl Acad Sci U S A 101:12543–12548

Reubinoff BE, Itsykson P, Turetsky T, Pera MF, Reinhartz E, Itzik A, Ben-Hur T (2001) Neural progenitors from human embryonic stem cells. Nat Biotechnol 19:1134–1140

Sánchez-Pernaute R, Studer L, Bankiewicz KS, Major EO, McKay RDG (2001) In vitro generation and transplantation of precursor-derived human dopamine neurons. J Neurosci Res 65:284–288

Schierle GS, Hansson O, Leist M, Nicotera P, Widner H, Brundin P (1999) Caspase inhibition reduces apoptosis and increases survival of nigral transplants. Nat Med 5:97–100

Schulz TC, Noggle SA, Palmarini GM, Weiler DA, Lyons IG, Pensa KA, Meedeniya AC, Davidson BP, Lambert NA, Condie BG (2004) Differentiation of human embryonic stem cells to dopaminergic neurons in serum-free suspension culture. Stem Cells 22:1218–1238

Storch A, Paul G, Csete M, Boehm BO, Carvey PM, Kupsch A, Schwarz J (2001) Long-term proliferation and dopaminergic differentiation of human mesencephalic neural precursor cells. Exp Neurol 170:317–325

Studer L, Tabar V, McKay RD (1998) Transplantation of expanded mesencephalic precursors leads to recovery in parkinsonian rats. Nat Neurosci 1:290–295

Svendsen CN, Caldwell MA, Shen J, ter Borg MG, Rosser AE, Tyers P, Karmiol S, Dunnett SB (1997) Long-term survival of human central nervous system progenitor cells transplanted into a rat model of Parkinson's disease. Exp Neurol 148:135–146

Takagi Y, Takahashi J, Saiki H, Morizane A, Hayashi T, Kishi Y, Fukuda H, Okamoto Y, Koyanagi M, Ideguchi M, Hayashi H, Imazato T, Kawasaki H, Suemori H, Omachi S, Iida H, Itoh N, Nakatsuji N, Sasai Y, Hashimoto N (2005) Dopaminergic neurons generated from monkey embryonic stem cells function in a Parkinson primate model. J Clin Invest 115:102–109

Vescovi AL, Parati EA, Gritti A, Poulin P, Ferrario M, Wanke E, Frolichsthal-Schoeller P, Cova L, Arcellana-Panlilio M, Colombo A, Galli R (1999) Isolation and cloning of multipotential stem cells from the embryonic human CNS and establishment of trans-plantable human neural stem cell lines by epigenetic stimulation. Exp Neurol 156:71–83

Wakayama T, Tabar V, Rodriguez I, Perry AC, Studer L, Mombaerts P (2001) Differentiation of embryonic stem cell lines generated from adult somatic cells by nuclear transfer. Science 292:740–743

Widner H, Tetrud J, Rehncrona S, Snow B, Brundin P, Gustavii B, Björklund A, Lindvall O, Langston JW (1992) Bilateral fetal mesencephalic grafting in two patients with parkinsonism induced by 1-methyl-4-phenyl-1, 2, 3, 6-tetrahydropyridine (MPTP). N Engl J Med 327:1556–1563

Wood HB, Episkopou V (1999) Comparative expression of the mouse Sox1, Sox2 and Sox3 genes from pre-gastrulation to early somite stages. Mech Dev 86:197–201

Ying QL, Stavridis M, Griffiths D, Li M, Smith A (2003) Conversion of embryonic stem cells into neuroectodermal precursors in adherent monocultures. Nat Biotechnol 21:183–186

Zeng X, Cai J, Chen J, Luo Y, You ZB, Fotter E, Wang Y, Harvey B, Miura T, Backman C, Chen GJ, Rao MS, Freed WJ (2004) Dopaminergic differentiation of human embryonic stem cells. Stem Cells 22:925–940

Zhang SC, Wernig M, Duncan ID, Brustle O, Thomson JA (2001) In vitro differentiation of transplantable neural precursors from human embryonic stem cells. Nat Biotechnol 19:1129–1133

Ernst Schering Research Foundation Workshop

Editors: Günter Stock
    Monika Lessl

Vol. 1 (1991): Bioscience ⇌ Societly Workshop Report
Editors: D.J. Roy, B.E. Wynne, R.W. Old

Vol. 2 (1991): Round Table Discussion on Bioscience ⇌ Society
Editor: J.J. Cherfas

Vol. 3 (1991): Excitatory Amino Acids and Second Messenger Systems
Editors: V.I. Teichberg, L. Turski

Vol. 4 (1992): Spermatogenesis – Fertilization – Contraception
Editors: E. Nieschlag, U.-F. Habenicht

Vol. 5 (1992): Sex Steroids and the Cardiovascular System
Editors: P. Ramwell, G. Rubanyi, E. Schillinger

Vol. 6 (1993): Transgenic Animals as Model Systems for Human Diseases
Editors: E.F. Wagner, F. Theuring

Vol. 7 (1993): Basic Mechanisms Controlling Term and Preterm Birth
Editors: K. Chwalisz, R.E. Garfield

Vol. 8 (1994): Health Care 2010
Editors: C. Bezold, K. Knabner

Vol. 9 (1994): Sex Steroids and Bone
Editors: R. Ziegler, J. Pfeilschifter, M. Bräutigam

Vol. 10 (1994): Nongenotoxic Carcinogenesis
Editors: A. Cockburn, L. Smith

Vol. 11 (1994): Cell Culture in Pharmaceutical Research
Editors: N.E. Fusenig, H. Graf

Vol. 12 (1994): Interactions Between Adjuvants, Agrochemical
and Target Organisms
Editors: P.J. Holloway, R.T. Rees, D. Stock

Vol. 13 (1994): Assessment of the Use of Single Cytochrome
P450 Enzymes in Drug Research
Editors: M.R. Waterman, M. Hildebrand

Vol. 14 (1995): Apoptosis in Hormone-Dependent Cancers
Editors: M. Tenniswood, H. Michna

Vol. 15 (1995): Computer Aided Drug Design in Industrial Research
Editors: E.C. Herrmann, R. Franke

Vol. 16 (1995): Organ-Selective Actions of Steroid Hormones
Editors: D.T. Baird, G. Schütz, R. Krattenmacher

Vol. 17 (1996): Alzheimer's Disease
Editors: J.D. Turner, K. Beyreuther, F. Theuring

Vol. 18 (1997): The Endometrium as a Target for Contraception
Editors: H.M. Beier, M.J.K. Harper, K. Chwalisz

Vol. 19 (1997): EGF Receptor in Tumor Growth and Progression
Editors: R.B. Lichtner, R.N. Harkins

Vol. 20 (1997): Cellular Therapy
Editors: H. Wekerle, H. Graf, J.D. Turner

Vol. 21 (1997): Nitric Oxide, Cytochromes P 450,
and Sexual Steroid Hormones
Editors: J.R. Lancaster, J.F. Parkinson

Vol. 22 (1997): Impact of Molecular Biology
and New Technical Developments in Diagnostic Imaging
Editors: W. Semmler, M. Schwaiger

Vol. 23 (1998): Excitatory Amino Acids
Editors: P.H. Seeburg, I. Bresink, L. Turski

Vol. 24 (1998): Molecular Basis of Sex Hormone Receptor Function
Editors: H. Gronemeyer, U. Fuhrmann, K. Parczyk

Vol. 25 (1998): Novel Approaches to Treatment of Osteoporosis
Editors: R.G.G. Russell, T.M. Skerry, U. Kollenkirchen

Vol. 26 (1998): Recent Trends in Molecular Recognition
Editors: F. Diederich, H. Künzer

Vol. 27 (1998): Gene Therapy
Editors: R.E. Sobol, K.J. Scanlon, E. Nestaas, T. Strohmeyer

Vol. 28 (1999): Therapeutic Angiogenesis
Editors: J.A. Dormandy, W.P. Dole, G.M. Rubanyi

Vol. 29 (2000): Of Fish, Fly, Worm and Man
Editors: C. Nüsslein-Volhard, J. Krätzschmar

Vol. 30 (2000): Therapeutic Vaccination Therapy
Editors: P. Walden, W. Sterry, H. Hennekes

Vol. 31 (2000): Advances in Eicosanoid Research
Editors: C.N. Serhan, H.D. Perez

Vol. 32 (2000): The Role of Natural Products in Drug Discovery
Editors: J. Mulzer, R. Bohlmann

Vol. 33 (2001): Stem Cells from Cord Blood, In Utero Stem Cell Development, and Transplantation-Inclusive Gene Therapy
Editors: W. Holzgreve, M. Lessl

Vol. 34 (2001): Data Mining in Structural Biology
Editors: I. Schlichting, U. Egner

Vol. 35 (2002): Stem Cell Transplantation and Tissue Engineering
Editors: A. Haverich, H. Graf

Vol. 36 (2002): The Human Genome
Editors: A. Rosenthal, L. Vakalopoulou

Vol. 37 (2002): Pharmacokinetic Challenges in Drug Discovery
Editors: O. Pelkonen, A. Baumann, A. Reichel

Vol. 38 (2002): Bioinformatics and Genome Analysis
Editors: H.-W. Mewes, B. Weiss, H. Seidel

Vol. 39 (2002): Neuroinflammation – From Bench to Bedside
Editors: H. Kettenmann, G.A. Burton, U. Moenning

Vol. 40 (2002): Recent Advances in Glucocorticoid Receptor Action
Editors: A. Cato, H. Schaecke, K. Asadullah

Vol. 41 (2002): The Future of the Oocyte
Editors: J. Eppig, C. Hegele-Hartung

Vol. 42 (2003): Small Molecule-Protein Interaction
Editors: H. Waldmann, M. Koppitz

Vol. 43 (2003): Human Gene Therapy:
Present Opportunities and Future Trends
Editors: G.M. Rubanyi, S. Ylä-Herttuala

Vol. 44 (2004): Leucocyte Trafficking:
The Role of Fucosyltransferases and Selectins
Editors: A. Hamann, K. Asadullah, A. Schottelius

Vol. 45 (2004): Chemokine Roles in Immunoregulation and Disease
Editors: P.M. Murphy, R. Horuk

Vol. 46 (2004): New Molecular Mechanisms of Estrogen Action
and Their Impact on Future Perspectives in Estrogen Therapy
Editors: K.S. Korach, A. Hillisch, K.H. Fritzemeier

Vol. 47 (2004): Neuroinflammation in Stroke
Editors: U. Dirnagl, B. Elger

Vol. 48 (2004): From Morphological Imaging to Molecular Targeting
Editors: M. Schwaiger, L. Dinkelborg, H. Schweinfurth

Vol. 49 (2004): Molecular Imaging
Editors: A.A. Bogdanov, K. Licha

Vol. 50 (2005): Animal Models of T Cell-Mediated Skin Diseases
Editors: T. Zollner, H. Renz, K. Asadullah

Vol. 51 (2005): Biocombinatorial Approaches for Drug Finding
Editors: W. Wohlleben, T. Spellig, B. Müller-Tiemann

Vol. 52 (2005): New Mechanisms for Tissue-Selective Estrogen-Free Contraception
Editors: H.B. Croxatto, R. Schürmann, U. Fuhrmann, I. Schellschmidt

Vol. 53 (2005): Opportunities and Challenges of the Therapies Targeting CNS Regeneration
Editors: D. Perez, B. Mitrovic, A. Baron Van Evercooren

Vol. 54 (2005): The Promises and Challenges of Regenerative Medicine
Editors: J. Morser, S.I. Nishikawa

Vol. 55 (2006): Chronic Viral and Inflammatory Cardiomyopathy
Editors: H.-P. Schultheiss, J.-F. Kapp

Vol. 56 (2006): Cytokines as Potential Therapeutic Target for Inflammatory Skin Diseases
Editors: R. Numerof, C.A. Dinarello, K. Asadullah

Vol. 57 (2006): The Histone Code and Beyond
Editors: S.L. Berger, O. Nakanishi, B. Haendler

Vol. 58 (2006): Chemical Genomics
Editors: S. Jaroch, H. Weinmann

Vol. 59 (2006): Appropriate Dose Selection – How to Optimize Clinical Drug Development
Editors: J. Venitz, W. Sittner

Vol. 60 (2006): Stem Cells in Reproduction and in the Brain
Editors: J. Morser, S.-I. Nishikawa, H.R. Schöler

Supplement 1 (1994): Molecular and Cellular Endocrinology of the Testis
Editors: G. Verhoeven, U.-F. Habenicht

Supplement 2 (1997): Signal Transduction in Testicular Cells
Editors: V. Hansson, F.O. Levy, K. Taskén

Supplement 3 (1998): Testicular Function:
From Gene Expression to Genetic Manipulation
Editors: M. Stefanini, C. Boitani, M. Galdieri, R. Geremia,F. Palombi

Supplement 4 (2000): Hormone Replacement Therapy
and Osteoporosis
Editors: J. Kato, H. Minaguchi, Y. Nishino

Supplement 5 (1999): Interferon: The Dawn of Recombinant
Protein Drugs
Editors: J. Lindenmann, W.D. Schleuning

Supplement 6 (2000): Testis, Epididymis and Technologies
in the Year 2000
Editors: B. Jégou, C. Pineau, J. Saez

Supplement 7 (2001): New Concepts in Pathology
and Treatment of Autoimmune Disorders
Editors: P. Pozzilli, C. Pozzilli, J.-F. Kapp

Supplement 8 (2001): New Pharmacological Approaches
to Reproductive Health and Healthy Ageing
Editors: W.-K. Raff, M.F. Fathalla, F. Saad

Supplement 9 (2002): Testicular Tangrams
Editors: F.F.G. Rommerts, K.J. Teerds

Supplement 10 (2002): Die Architektur des Lebens
Editors: G. Stock, M. Lessl

Supplement 11 (2005): Regenerative and Cell Therapy
Editors: A. Keating, K. Dicke, N. Gorin, R. Weber, H. Graf

Supplement 12 (2005): Von der Wahrnehmung zur Erkenntnis –
From Perception to Understanding
Editors: M. Lessl, J. Mittelstraß

Printing: Krips bv, Meppel
Binding: Stürtz, Würzburg